T0260338

Exploring Modeling with Data and Differential Equations Using R

Exploring Modeling with Data and Differential Equations Using R provides a unique introduction to differential equations with applications to the biological and other natural sciences. Additionally, model parameterization and simulation of stochastic differential equations are explored, providing additional tools for model analysis and evaluation. This unified framework sits "at the intersection" of different mathematical subject areas, data science, statistics, and the natural sciences. The text throughout emphasizes data science workflows using the R statistical software program and the tidyverse constellation of packages. Only knowledge of calculus is needed; the text's integrated framework is a stepping stone for further advanced study in mathematics or as a comprehensive introduction to modeling for quantitative natural scientists.

The text will introduce you to:

- modeling with systems of differential equations and developing analytical, computational, and visual solution techniques.
- the R programming language, the tidyverse syntax, and developing data science workflows.
- qualitative techniques to analyze a system of differential equations.
- data assimilation techniques (simple linear regression, likelihood or cost functions, and Markov Chain, Monte Carlo Parameter Estimation) to parameterize models from data.
- simulating and evaluating outputs for stochastic differential equation models.

An associated R package provides a framework for computation and visualization of results.

John Zobitz is a Professor of Mathematics and Data Science at Augsburg University in Minneapolis, Minnesota. His scholarship in environmental data science includes ecosystem models parameterized with datasets from environmental observation networks. He is a member of the Mathematical Association of America (MAA) and previous president of the North Central Section of the MAA. He has served on the editorial board of *MAA Notes*. He was a recipient of the Fulbright-Saastamoinen Foundation Grant in Health and Environmental Sciences at the University of Eastern Finland in Kuopio, Finland. In addition, he is an affiliated member of the Ecological Forecasting Network and regularly taught at Fluxcourse, an annual summer course for measurements and modeling of ecosystem biogeochemical fluxes.

Exploring Modeling with Data and Differential Equations Using R

John M. Zobitz

CRC Press
Taylor & Francis Group
Boca Raton London New York

CRC Press is an imprint of the
Taylor & Francis Group, an **informa** business

A CHAPMAN & HALL BOOK

First edition published 2023
by CRC Press
6000 Broken Sound Parkway NW, Suite 300, Boca Raton, FL 33487-2742

and by CRC Press
4 Park Square, Milton Park, Abingdon, Oxon, OX14 4RN

CRC Press is an imprint of Taylor & Francis Group, LLC

Library of Congress Cataloging-in-Publication Data

Names: Zobitz, John M., author.
Title: Exploring modeling with data and differential equations using R /
John M. Zobitz.
Description: First edition. | Boca Raton, FL : CRC Press, 2023. | Includes
bibliographical references and index.
Identifiers: LCCN 2022022121 (print) | LCCN 2022022122 (ebook) | ISBN
9781032259482 (hbk) | ISBN 9781032261812 (pbk) | ISBN 9781003286974
(ebk)
Subjects: LCSH: Biological models--Data processing. | Biological
models--Mathematical models. | Differential equations. | R (Computer
program language)
Classification: LCC QH324.8 .Z63 2023 (print) | LCC QH324.8 (ebook) | DDC
570.285--dc23/eng/20220831
LC record available at https://lccn.loc.gov/2022022121
LC ebook record available at https://lccn.loc.gov/2022022122

ISBN: 978-1-032-25948-2 (hbk)
ISBN: 978-1-032-26181-2 (pbk)
ISBN: 978-1-003-28697-4 (ebk)

DOI: 10.1201/9781003286974

Typeset in Latin Modern
by KnowledgeWorks Global Ltd.

Publisher's note: This book has been prepared from camera-ready copy provided by the authors.

To my parents Joan and Francis,

who gave me strong roots from which to grow.

To my wife Shannon,

who provides me the support to keep my trunk from breaking.

To my children Colin, Grant, Phoebe,

who are the branches that support the emerald green leaves.

Kiitos. Amor a todos.

Contents

List of Figures

Welcome

This book is written for you, the student learning about modeling and differential equations. Perhaps you first encountered models, differential equations, and better yet, building plausible models from data in your calculus course.

This book sits "at the intersection" of several different mathematics courses: differential equations, linear algebra, statistics, calculus, data science - as well as the partner disciplines of biology, chemistry, physics, business, and economics. An important idea is one of *transference* where a differential equation model applied in one context can also be applied (perhaps with different variable names) in a separate context.

I intentionally emphasize models from biology and the environmental sciences, but throughout the text you can find examples from the other disciplines. In some cases I've created homework exercises based on sources that I have found useful for teaching (denoted with "Inspired by . . . "). I hope you see the connections of this content to your own intended major.

This book is divided into 4 parts:

1. Models with differential equations
2. Parameterizing models with data
3. Stability analysis for differential equations
4. Stochastic differential equations

You may notice the interwoven structure for this book: models are introduced first, followed by data analysis and parameter estimation, returning back to analyzing models, and ending with simulating random (stochastic) models.

Unsure what about all these topics mean? Do not worry! The topics are presented with a "modeling first" paradigm that first introduces models, and equally important, explains how data are used to inform a model. This "conversation" between models and data are important to help build plausibility and confidence in a model. Stability analysis helps to solidify the connection between models and parameters (which may change the underlying dynamical stability). Finally the notion of *randomness* is extended with the introduction of stochastic differential equations.

Unifying all of these approaches is the idea of developing workflows for analysis, visualization results, and interpreting any results in the context of the problem.

Computational code

This book makes heavy use of the R programming language, and unabashedly develops programming principles using the tidyverse syntax and programming approach. This is intentional to facilitate direct connections to courses in introductory data science or data visualization. Throughout my years learning (and teaching) different programming languages, I have found R to be the most versatile and adaptable. The tidyverse syntax, in my opinion, has transformed my own thinking about sustainable computation and modeling processes - and I hope it does for you as well.

There is a companion R package available called demodelr to run programs and functions in the text (Zobitz 2022). Instructions to install this package are given in Chapter 2. The minimum version of R used was Version 4.0.2 (2020-06-22) (R Core Team 2021) and RStudio is Version 1.4.1717 (RStudio Team 2020).

The demodelr package uses the following R packages:

- tidyverse (and the associated packages) (Version 1.3.1) (Wickham et al. 2019)
- GGally (Version 2.1.2) (Schloerke et al. 2021)
- formula.tools (Version 1.7.1) (Brown 2018)
- expm (Version 0.999-6) (Goulet et al. 2021)

Questions? Comments? Issues?

Any errors or omissions are of my own accord, so please contact me at zobitz@augsburg.edu. Feel free to file an issue with the demodelr package to my github.[1]

About the cover

The photo on the back cover was taken by Shannon Zobitz during a hike at Orinoro Gorge[2] in Finland. The photo is indicative of several things: (1) the journey ahead as you commence learning about modeling, differential

---

[1] https://github.com/jmzobitz/ModelingWithR/issues
[2] https://visitleppavirta.fi/en/service/orinoro-gorge

equations, and R, (2) the occasional roots in the path that may cause you to stumble (such as coding errors). Everyone makes them, so you are in good company. (3) the yellow markings on the trees indicate the way forward. The vector field image on the cover is an example of a spiral node, indicating my hope that the knowledge contained here spirals out and informs your future endeavors. May this textbook be the guide for you as you progress over the hill and onward. Let's get started!

Acknowledgments

This book has been developed over the course of several years in a variety of places: two continents, between meetings, in the early mornings, at coffee shops, or while waiting for practices to end. Special thanks are to the following:

- **Augsburg University:** You have been my professional home for over a decade and given me the space and support to be intellectually creative in my teaching and scholarship. Special thanks to my Mathematics, Statistics, and Computer Science Department colleagues - it is a joy to work with all of you.

- **Augsburg University students:** Thank you for your interest and engagement in this topic, allowing me to test ideas in an upper division course titled (wait for it ...) *Modeling and Differential Equations in the Biological and Natural Sciences*. While the course title is a mouthful, you provided concise, honest, and insightful feedback, shaping this text. I am forever indebted to you. Kiitos to students in the Fall 2019 and 2021 courses.

- **My family:** Shannon, Colin, Grant, and Phoebe for humoring me (and my occasional grumpiness) while this project has been completed.

- **Taylor & Francis:** Thank you for your confidence in me with this project, and to my editor Lara Spieker for shepherding the project and Robin Lloyd Starkes and her team for their careful copyediting.

Part I

Models with Differential Equations

1

Models of Rates with Data

1.1 Rates of change in the world: a model is born

This book focuses on understanding *rates of change* and their application to modeling real-world phenomena with contexts from the natural sciences. Additionally, this book emphasizes *using* equations with data, building both competence and confidence to construct and evaluate a mathematical model with data. Perhaps these emphases are different from when you analyzed rates of change in a calculus course; consider the following types of questions:

- If $y = xe^{-x}$, what is the derivative function $f'(x)$?
- What is the equation of the tangent line to $y = x^3 - x$ at $a = 1$?
- Where is the graph of $\sin(x)$ increasing at an increasing rate?
- If you release a ball from the top of a skyscraper 500 meters above the ground, what is its speed when it impacts the ground?
- What is the largest area that can be enclosed in a chicken coop with 100 feet of fencing, with one side being along a wall?

The first three questions do not appear to be connected in a real-world context in their framing - but the last two questions *do* have some context from real-world situations. The given context may reveal underlying assumptions or physical principles, which are the starting point to build a mathematical model. For the chicken coop problem, perhaps the next step is to use the assumed geometry (rectangle) with the 100 feet of fencing to develop a function for the area.

Maybe the context includes observational data and several different (perhaps conflicting) assumptions about the context at hand. For example, how does air resistance affect the ball's velocity? Would a circular chicken coop maximize the area more than a rectangular coop? For both of these cases, which model is the best one to approximate any observational data? The short answer: it depends. To understand why, let's take a look at another problem in context.

1.2   Modeling in context: the spread of a disease

Consider the data in Figure 1.1, which come from an Ebola outbreak[1] in Sierra Leone in 2014. (Data provided from Matthes (2021).) The vertical axis in Figure 1.1 represents Ebola *infections* over 2 years from initial monitoring in March 2014.

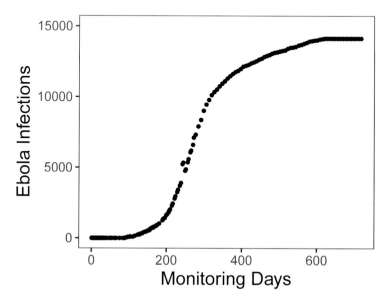

FIGURE 1.1 Infections from a 2014 Ebola outbreak in Sierra Leone, with the initial monitoring in March 2014.

Constructing a model from disease dynamics is part of the field of mathematical epidemiology.[2] Here we focus on person to person or population spread of Ebola. Other types of models could focus on the immune response within a single person - perhaps with a goal to design effective types of treatments to reduce the severity of infection. How we construct a mathematical model for this outbreak largely depends on the assumptions underlying the biological dynamics of disease transmission (which we will call the *infection rate*). Three plausible assumptions for the infection rate are the following:

1. The infection rate is proportional to the number of people infected.
2. The infection rate is proportional to the number of people **not** infected.
3. The infection rate is proportional to the number of infected people coming into contact with those not infected.

---

[1]https://www.cdc.gov/vhf/ebola/history/2014-2016-outbreak/index.html
[2]https://en.wikipedia.org/wiki/Mathematical_modelling_of_infectious_disease

Now let's explore how to translate these assumptions into a mathematical model. Since we are discussing *rates* of infection, this means we will need a *rate of change* or derivative. Let's use the letter I to represent the number of people that are infected.

1.2.1 Model 1: Infection rate proportional to number infected

The first assumption states that the infection rate is proportional to the number of people infected. Translated into an equation this would be the following:

$$\frac{dI}{dt} = kI \qquad (1.1)$$

Equation (1.1) is an example of a *differential equation*, which is just a mathematical equation with rates of change. In Equation (1.1) k is a proportionality constant or parameter, with units of time^{-1} for consistency.

The *solution* to a differential equation is a function $I(t)$. When we "solve" a differential equation we determine the family of functions consistent with our rate equation.[3] There are a lot of techniques to solve a differential equation; we will explore some in Chapter 7.

The proportionality constant or *parameter* k is important to understand the solution to Equation (1.1). Even though no numerical value for k is specified, you can always solve an equation without specifying the parameter. In some situations we may not be as concerned with the particular *value* of the parameter but rather its influence on the long-term behavior of the system (this is a key aspect of bifurcation theory described in Chapter 20). Otherwise we can use the collected data shown above with the given model to determine the value for k. This combination of a mathematical model with data is called *data assimilation* or *model-data fusion* (see Chapters 8-14).

How plausible is this first model? The first model assumes the rate of change (Equation (1.1)) gets larger as the number of infected people I increases. This reasoning certainly seems plausible: when there are so many people infected it can be hard to stay healthy! At some point the number of people who are *not* sick will reach zero, making the rate of infection zero (or no increase). In the case of Ebola or any other infectious disease, stringent public health measures would be enacted if the number of people infected became too large.[4] Following public health measures we would expect that the rate of infection would decrease and the number of infections to slow. So perhaps another model this can capture this "slowing down" of the infection rate is more plausible.

---

[3]You may be used to working with *algebraic equations* (e.g. solve $x^2 - 4 = 0$ for x) rather than differential equations. For algebraic equations the solution can be points (for our example, the solution to $x^2 - 4 = 0$ is $x = \pm 2$).

[4]The COVID-19 pandemic that began in 2020 is an example of the heroic efforts of public health officials.

1.2.2   Model 2: Infection rate proportional to number NOT infected

The second model considers the interaction between people who are sick (which we have denoted as I) and people who are *not* sick, which we will call S, or susceptible. Equation (1.2) is an example of a differential equation that models this interation:

$$\frac{dI}{dt} = kS \tag{1.2}$$

As with Equation (1.1) the parameter k represents an infection rate. We would expect that both I and S change in time as the infection occurs; for a finite population as more people get sick (I), that would mean that S would decrease. In effect, Model 2 should have *two* rates of change: one for I and one for S. Figure 1.2 shows a schematic of this process of infection.

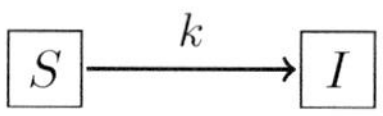

FIGURE 1.2 Schematic diagram for Model 2, showing that the rate of infection is proportional to the number of susceptible people S. Assuming a constant population size N, the differential equation for Model 2 is given by Equation 1.4.

There are three reasons why I like to use diagrams like Figure 1.2:

(1) Diagrams build a bridge between biological processes and mathematical models.
(2) Diagrams signal which rates (if any) can be conserved (more on this below).
(3) Diagrams help to identify assumed parameters (i.e. k in Figure 1.2).
(4) Diagrams suggest how to construct differential equations for this mathematical model. Figure 1.2 suggests a flow between the suspectible state S to the infected state I. So then the rate of change equation for S is $\frac{dS}{dt} = -kS$ (the parameter listed above the arrow in Figure 1.2). Equation (1.3) combines all this thinking and Equation (1.2) into the following coupled system of differential equations in Equation (1.3):

$$\begin{aligned} \frac{dS}{dt} &= -kS \\ \frac{dI}{dt} &= kS \end{aligned} \tag{1.3}$$

The solution to Equation (1.3) is functions $S(t)$ and $I(t)$ that evolve over time. We don't have the tools to determine the exact solutions for Equation (1.3) yet (we will study systems like these in Chapters 15-20). However something interesting occurs with Equation (1.3) when we add the rates $\dfrac{dS}{dt}$ and $\dfrac{dI}{dt}$ together (Equation (1.4)):

$$\frac{dS}{dt} + \frac{dI}{dt} = \frac{d}{dt}(S+I) = 0 \tag{1.4}$$

If a rate of change equals zero then the corresponding function is constant. In effect, Equation (1.4) means that the combined variable $S+I$ is constant, so we could say that $S+I = N$, where N is the total population size. The expression $S+I = N$ is an example of a conservation law for our system.[5] Figure 1.2 also suggests a conservation law because there are no additional arrows going into or from the variables S or I. Since $S = N - I$, Equation (1.3) can be re-written with a single equation (Equation (1.5)):

$$\frac{dI}{dt} = k(N - I) \tag{1.5}$$

Equation (1.5) also indicates limiting behavior for Model 2. As the number of infected people reaches N (the total population size), the values of $\dfrac{dI}{dt}$ approaches zero, meaning I doesn't change. Biologically this would suggest that eventually everyone in the population would get sick with the disease (assuming no one has any natural immunity). Equation (1.5) also has one caveat: if there are no infected people around ($I = 0$) *the disease can still be transmitted*, which might not make good biological sense. The next model (Model 3) tries to amend that shortcoming.

1.2.3 Model 3: Infection rate proportional to infected meeting not infected

Now consider a third model that rectifies some of the shortcomings of the second model (the second model rectified the shortcomings of the first model). The third model states that the rate of infection is due to those who are sick infecting those who are not sick. This scenario would also make some sense, as it focuses on the *transmission* of the disease between susceptibles and infected people. So if nobody is sick ($I = 0$) then the disease is not spread. Likewise if there are no susceptibles ($S = 0$), the disease is not spread as well.

In this case the diagram outlining the third model looks something like this:

---

[5] A finite population (meaning nobody can exit or enter the population) usually should have some type of conservation law.

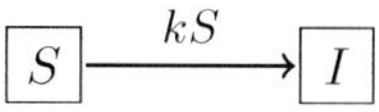

$$S \xrightarrow{\quad kS \quad} I$$

FIGURE 1.3 Schematic diagram for Model 3, showing that the rate of infection is proportional to the number of susceptible people S encountering an infected person I. Assuming a constant population size N, the differential equation for Model 3 is given by Equation 1.7.

Notice how in Figure 1.3 there is an additional variable S associated with k to show how the rate of infection depends on S. Equation (1.6) contains the differential equations that describe the scenario outlined in Figure 1.3:

$$\begin{aligned} \frac{dS}{dt} &= -kSI \\ \frac{dI}{dt} &= kSI \end{aligned} \tag{1.6}$$

Similar to Model 2 we can combine the two equations to yield a single differential equation (Equation (1.7)):

$$\frac{dI}{dt} = k \cdot I \cdot (N - I) \tag{1.7}$$

Equation (1.7) appears similar to Equation (1.5), doesn't it? However in Equation (1.7) notice the variable I outside the expression $(N - I)$. If $I = 0$, then there is no increase in infection (the rate is zero). If $I = N$ (the total population size) then there is no increase in the infection (the rate is zero as well). Model 3 seems to be more consistent with the biological reasoning for the spread of infection.

Let's compare all the rates for all three models together in Figure 1.4. Figure 1.4 has a lot to unpack, but we can use some of our understanding of rates of change in calculus to compare the three models. Notice how the sign of $\frac{dI}{dt}$ is always positive for Model 1, indicating that the solution (I) is always increasing. For Models 2 and 3, $\frac{dI}{dt}$ equals zero when $I = 10$, which also is the value for N After that case, $\frac{dI}{dt}$ turns negative, meaning that I is decreasing.

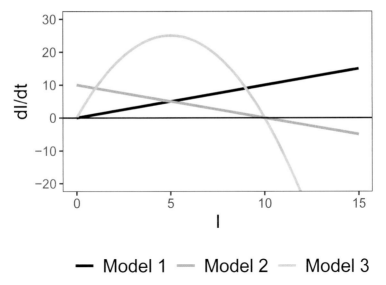

FIGURE 1.4 Comparing the rates of change for three models (Equation (1.1), Equation (1.5), and Equation (1.7), with $k = 1$ and $N = 10$).

In summary, examining the graphs of the rates can tell a lot about the *qualitative behavior* of a solution to a differential equation even without the solution.

1.3 Model solutions

Let's return back to possible solutions (in this case formulas for $I(t)$) for our models. Usually a differential equation also has a starting or an initial value (typically at $t = 0$) that actualizes the solution. When we state a differential equation with a starting value we have an **initial value problem**. We will represent that initial value as $I(0) = I_0$.

With that assumption, we can (and will solve later!) the following solutions for these models:

$$\text{Model 1 (Exponential): } I(t) = I_0 e^{kt}$$
$$\text{Model 2 (Saturating): } I(t) = N - (N - I_0)e^{-kt} \tag{1.8}$$
$$\text{Model 3 (Logistic): } I(t) = \frac{N \cdot I_0}{I_0 + (N - I_0)e^{-kt}}$$

Notice how I assigned the names to each model (Exponential, Saturating, and

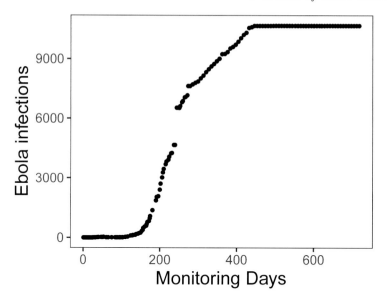

FIGURE 1.6 Infections from a 2014 Ebola outbreak in Liberia, with the initial monitoring in March 2014.

Exercise 1.2. Figure 1.6 shows the Ebola outbreak for the country of Liberia in 2014. If we were to apply the logistic model (Model 3) based on this graphic what would be your estimate for N?

Exercise 1.3. The general solutions for the saturating and the logistic models are:

$$\text{Saturating model: } I(t) = N - (N - I_0)e^{-kt}$$

$$\text{Logistic model: } I(t) = \frac{N \cdot I_0}{I_0 + (N - I_0)e^{-kt}}, \tag{1.10}$$

where I_0 is the initial number of people infected and N is the overall population size. Using the functions from Exercise 1.1 for both models, what are the values for N and I_0?

Exercise 1.4. The general solutions for the saturating and the logistic models are:

$$\text{Saturating model: } I(t) = N - (N - I_0)e^{-kt}$$

$$\text{Logistic model: } I(t) = \frac{N \cdot I_0}{I_0 + (N - I_0)e^{-kt}}, \tag{1.11}$$

where I_0 is the initial number of people infected and N is the overall population size. For both models carefully evaluate the limits to show $\lim_{t \to \infty} I(t) = N$. How do these limiting values compare to the steady-state values you found for Models 2 and 3 in Figure 1.5, where $N = 13600$?

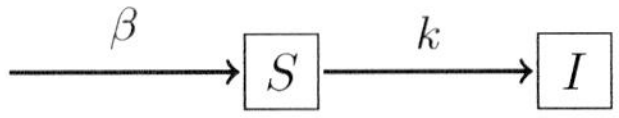

FIGURE 1.7 Schematic diagram for Exercise 1.5.

Exercise 1.5. Figure 1.7 shows a schematic diagram which is a variation on Figure 1.2. In this case people are entering the the susceptible population S at a rate β, so the population is not conserved. What is the coupled system of differential equations for this model?

Exercise 1.6. A model that describes the growth of sales of a product in response to advertising is the following:

$$\frac{dS}{dt} = .55\sqrt{1 - S} - S, \tag{1.12}$$

where S is the product's share of the market (scaled between 0 and 1) (Sethi 1983). Use this information to answer the following questions:

a. Make a plot of the function $f(S) = .55\sqrt{1 - S} - S.$ for $0 \le S \le 1$.
b. Interpret your plot to predict when the market share will be increasing and decreasing. At what value is $\dfrac{dS}{dt} = 0$? (This is called the *steady state* value.)
c. A second campaign has the following differential equation:

$$\frac{dS}{dt} = .2\sqrt{1 - S} - S \tag{1.13}$$

What is the steady state value and how does it compare to the previous one?

Exercise 1.7. A more general form of the advertising model is

$$\frac{dS}{dt} = r\sqrt{1 - S} - S, \tag{1.14}$$

where S is the product's share of the market (scaled between 0 and 1). The parameter r is related to the effectiveness of the advertising (between 0 and 1).

a. Solve $\dfrac{dS}{dt} = r\sqrt{1 - S} - S$ for the steady state value (where $\dfrac{dS}{dt} = 0$). Your final answer should be expressed as a function $S(r)$ - for which you will need to use the quadratic formula.
b. Make a plot of the steady state value as a function of r, where $0 \le r \le 1$.
c. Based on your plot, what can you conclude about the steady state value as the effectiveness of the advertising increases?

Exercise 1.8. A common saying is "you are what you eat." An equation that relates an organism's nutrient content (denoted as y) to the nutrient content of food or resource (denoted as x) is given by:

$$y = cx^{1/\theta}, \tag{1.15}$$

where θ and c are both constants. Units on x and y are expressed as a proportion of a given nutrient (such as nitrogen or carbon). For example, when $c = 1$ and $\theta = 1$ the function is $y = x$. In this case the point $(0.05, 0.05)$ would say that nutrient composition for the organism and resource would be the same.

a. Now assume that $c = 1$. How does the nutrient content of the organism compare to the resource when $\theta = 2$? Draw a sample curve and interpret it, contrasting it to when $\theta = 1$.
b. Now assume that $c = 1$. How does the nutrient content of the organism compare to the resource when $\theta = 5$? Draw a sample curve and interpret it, contrasting this curve to the previous two.
c. What do you think will happen when $\theta \to \infty$? Draw some sample curves to help illustrate your findings.

Exercise 1.9. A model for the outbreak of a cold virus assumes that the rate people get infected is proportional to infected people contacting susceptible people, as with Model 3 (the Logistic model). However people who are infected can also recover and become susceptible again with rate α. Construct a diagram similar to Figure 1.3 for this scenario and also write down what you think the system of differential equations would be.

Exercise 1.10. A model for the outbreak of the flu assumes that the rate people get infected is proportional to infected people contacting susceptible people, as in Model 3. However people also recover from the flu, denoted with the variable R. Assume that the rate of recovery is proportional to the number of infected people with parameter β. Construct a diagram similar to Figure 1.3 for this scenario and also write down what you think the system of differential equations would be.

Exercise 1.11. (Inspired by Hugo van den Berg (2011)) Organisms that live in a saline environment biochemically maintain the amount of salt in their blood stream. An equation that represents the level of S in the blood is the following:

$$\frac{dS}{dt} = I + p \cdot (W - S), \tag{1.16}$$

where the parameter I represents the active uptake of salt, p is the permeability of the skin, and W is the salinity in the water. Use this information to answer the following questions:

a. What is that value of S at *steady state*, or when $\dfrac{dS}{dt} = 0$? Your final answer should be a function $S(I, p, W)$.

b. With the steady state solution, what parameters (I, p, or W) cause the steady state value S to increase?

Exercise 1.12. (Inspired by Logan and Wolesensky (2009)) The immigration rate of bird species (species per time) from a mainland to an offshore island is $I_m \cdot (1 - S/P)$, where I_m is the maximum immigration rate, P is the size of the source pool of species on the mainland, and S is the number of species already occupying the island. Additionally the extinction rate is $E \cdot S/P$, where E is the maximum extinction rate. The growth rate of the number of species on the island is the immigration rate minus the extinction rate.

a. Make representative plots of the immigration and the extinction rates as a function of S. You may set I_m, P, and E all equal to 1.

b. Determine the number of species for which the net growth rate is zero, or the number of species is in equilibrium. Express your answer as S as a function of I_m, P, and E.

c. Suppose that two islands of the same size are at different distances from the mainland. Birds arrive from the source pool and they have the same extinction rate on each island. However the maximum immigration rate is larger for the island farther away. Which of the two islands will have the larger number of species at equilibrium?

Exercise 1.13. (Inspired by Logan and Wolesensky (2009)) Assume that an animal assimilates nutrients at a rate R proportional to its surface area. Also assume that it uses nutrients at a rate proportional to its volume. You may assume that the size of the animal is implicitly a function of the nutrient intake and usage, so $R = k_A A - k_V V$, where k_A and k_V are constants, A is the surface area, and V the volume. Determine expressions for the size of the animal if its intake and use rates were in balance (meaning R is set to zero), assuming the animal is the following shapes:

a. A sphere (assume size is measured with radius r) *Note:* first determine the geometric formulas for surface area and volume.

b. A cube (assume size is measured with length l)

2

Introduction to R

The primary tools we will use to analyze models for this book are R (R Core Team 2021) and RStudio (RStudio Team 2020).[1] These programs are powerful ones to learn! Admittedly learning a new software may be challenging; however I think it is worth it. With R you will have enormous flexibility to efficiently utilize data, design effective visualizations, and process statistical models. Let's get started!

2.1 R and RStudio

First let's talk terminology. The program RStudio is called an *Integrated Development Environment* for the statistical software language R.

To get both R and RStudio requires two separate downloads and files, which can be found here:

- R: https://cran.r-project.org/mirrors.html (You need to select a location to download from; choose any one that is geographically close to you.)
- RStudio: https://www.rstudio.com/products/rstudio/download/

2.1.1 Why do we have two programs?

Think of R as your basic program - this is the engine that does the computation. RStudio is a program where you can see everything you are working on in one place. Figure 2.1 shows an example of a typical RStudio workspace:

---

[1]In July 2022 RStudio announced it was changing its name to posit (https://www.rstudio.com/blog/rstudio-is-becoming-posit/). At this time of this writing posit was not available, although I expect its look and feel will be very similar to RStudio.

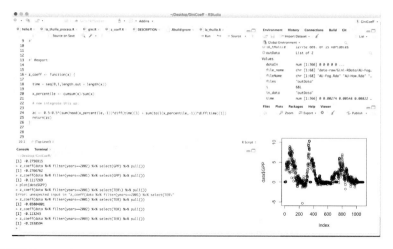

FIGURE 2.1 A sample `RStudio` workspace from one of my projects.

There are 4 key panels that I work with, clockwise from the top:

- The **source** window is in the upper left - notice how those have different tabs associated with them. You can have multiple source files that you can toggle between. For the moment think of these as commands that you will want to send to R.
- The **environment** and **history** pane - these tables allow you to see what variables are stored locally in your environment, or the history of commands.
- The **files** or **plots** pane (a simple plot I was working on is shown currently), but you can toggle between the tabs. The files tab shows the files in the current `Rstudio` project directory.
- Finally, the **console** pane is the place where R works and runs commands. You can type in there directly; otherwise we will also just "send" commands from the source down to the console.

Now we are ready to work with R and RStudio!

2.2   First steps: getting acquainted with R

Open up `RStudio`. The first task is to create a project file. A project is a central place to organize your materials. If you have previous experience with R you may be familiar with how the program is picky about its working directory - or the location on the computer where computations, files, and data are currently saved. Creating a project file is an easy way to avoid some of that fussiness. Here are the steps to accomplish this:

1. In RStudio select "File" then "New Project."
2. Next select the first option "New Directory" in the window - this will create a new folder on your computer.
3. At the next window choose "New Directory" or "Existing Directory." Depending on the option you choose, you will have some choice as to where you want to place this project.
4. Name the project as you like.
5. Click the "Create Project" button.

It might be helpful to think of a project file as a physical folder where you store papers that have something in common (such as class notes). When you want to work on the project, you open up your folder (and similarly close your project when you are done). At the point of return, you can re-open your folder and pick up where you left off. An RStudio project is similar in that regard as well.

2.2.1 Working with R

Our next step: how does R compute something? For example if we wanted to compute 4+9 we could type this command in the R console (lower left) window.[2] Try this out now:

1. In the console type 4+9
2. Then hit enter (or return)
3. Is the result 13?

Success! While this workflow is helpful for a single expression, making use of script files (.R file) can help run multiple steps of code at once. Script files are located in the upper left hand corner of your RStudio window - or the source window. (You may not have anything there when you start working on a project, so let's create one.)

1. In RStudio select "File" then "New File"
2. Next select the first option "New Script"
3. A new window called "UntitledX" should appear, where X is a number. You are set to go![3]

Source files allow you to type in R code and then evaluate, which is sometimes called "sending a command to the console" - or moving an R statement from the source window to the console. Working with a script file allows you to fix any coding errors more quickly and then re-run your code rather than re-type everything. Let's practice this.Type 4+9 in the script file. To evaluate this statement you have several options:

---

[2]I know you know the result is 13, but this is an illustrative example.

[3]**Pro tip:** There are shortcuts to creating a new file: Ctrl+Shift+N (Windows and Linux) or Command+Shift+N (Mac).

- too ambiguous.) I also used snake case to string together multiple words. In practice you can use snake case, or alphabetic cases (`myResult`) or even `my.result` (although that may not be preferred practice in the long run). However, if you name variables as `my-result` it looks like subtraction between variables `my` and `result`. I try to follow the tidyverse style guide[7] whenever possible.

Once we have defined a variable, we can compute with it. For example `10*my_result` should yield 130. Cool, no?

Sequences defined as vectors are another useful construction. In `R` this is done with the `seq` function along with additional information such as the starting value, ending value, and step size. As an example, let's define a sequence, spaced from 0 to 5 with spacing of 0.05 and then store this sequence as variable called `my_sequence`:

```
my_sequence <- seq(from = 0, to = 5, by = 0.05)
```

The format for the function `seq` is `seq(from=start,to=end,by=step_size)`. The `seq` command is a pretty flexible - there are alternative ways you can generate a sequence by specifying the starting and the end values along with the number of points. If you want to know more about `seq` you can always use `?` followed by the command - that will bring up the help values:

```
?seq
```

Once you get more comfortable with syntax in `R`, you will see that `seq(0,5,0.5)` gives the same result as `seq(from=0,to=5,by=0.05)`, but it is helpful to write your code *so that you can understand what it does.*[8]

2.4.2   Data frames

A key structure in `R` is that of a data frame, which allows different types of data to be collected together. A data frame is like a spreadsheet where each column is a value and each row a value (much like you would find in a spreadsheet). As an example, a data frame may list values for solutions to a differential equation, like we did with our three infection models in Chapter 1 (Table 2.1).

Data frames are an example of *tidy* data, where each row is an observation, each column a variable (which can be quantitative or categorical). There are several different ways to define a data frame in `R`. I am going to rely on the approach utilized by the `tidyverse`, which defines data frames as `tibbles`[9].

As an example, the following code defines a data frame that computes the quadratic function $y = 3x^2 - 2x$ for $-5 \leq x \leq 2$.

---

[7]https://style.tidyverse.org/
[8]I believe that code should be built for humans, not computers; see Wilson et al. (2014) for more information.
[9]https://tibble.tidyverse.org/

TABLE 2.1 Sample model solutions for an exponential, saturating, or logistic differential equation

time	model_1	model_2	model_3
0	250	250	250
1	258	645	257
2	265	1027	265
3	274	1399	273
4	282	1760	281

```
x <- seq(from = -5, to = 2, by = 0.05)
y <- 3 * x^2 - 2 * x

my_data <- tibble(
  x = x,
  y = y
) # Notice how x and y are specifically defined
```

Notice that the data frame `my_data` uses the column (variable) names of `x` and `y`. You could have also used `tibble(x,y)`, but it is helpful to name the columns in the way that you would like them to be named.

In addition to defining a data frame, R also contains several datasets in memory. In fact to see all the datasets, type `data()` at the console. Packages may also have datasets bundled with them. If you want to see the datasets for the `demodelr` package, you would type `data(package = "demodelr")` at the console.

2.4.3 Reading in datasets

Another R skill is importing data into R. Data come in several different types of formats, but one of the more versatile ones is a csv (**c**omma **s**eparated **v**alues) file. A csv file is a simplified version of an Excel or Google spreadsheet.[10] To read in the file you will use the command `read_csv` (part of the `readr` package in the `tidyverse`). The `read_csv` command which has the following structure, where FILENAME refers to the location of the file on your computer:

```
in_data <- read_csv(FILENAME)
```

For example the following code would read in a csv file of Ebola data located in the project directory:

---

[10]While the following steps focus on csv files, R can read in Excel files with the `readxl` package (https://readxl.tidyverse.org/) or Google sheets with the `googlesheets4` package (https://googlesheets4.tidyverse.org/).

```
ebola <- read_csv("ebola.csv")
```

Notice the quotes around the FILENAME.[11] The command `read_csv` is part of the `tidyverse`, but the function `read.csv` uses base R. They operate a little differently, but this book will use the `read_csv` command.

2.5   Visualization with R

Now we are ready to begin visualizing data frames. Two types of plots that we will need to make will be a scatter plot and a line plot. We are going to consider both of these separately, with examples that you should be able to customize.

2.5.1   Making a scatterplot

One dataset we have is the weight of a dog over time, adapted from this referenced website.[12] The data frame we will use is called `wilson` and is part of the `demodelr` library. You can also explore the documentation for this dataset by typing `?wilson` at the console. The `wilson` dataset has two variables here: $D =$ the age of the dog in days and $W =$ the weight of the dog in pounds. I have the data loaded into the `demodelr` package, which you can investigate by typing the following at the command line:

```
glimpse(wilson)
```

Notice that this data frame has two variables: `days` and `weight`. To make a scatter plot of these data we are going to use the command `ggplot` in Figure 2.2:

```
ggplot(data = wilson) +
  geom_point(aes(x = days, y = weight)) +
  labs(
    x = "Days since birth",
    y = "Weight (pounds)"
  )
```

---

[11]**Pro tips:** It is helpful to make a subfolder of your R project called data, where all csv files are stored. Then if you have the data files in the data folder, in RStudio you can type "data" and it may start to autocomplete - this is handy.

[12]http://bscheng.com/2014/05/07/modeling-logistic-growth-data-in-r/

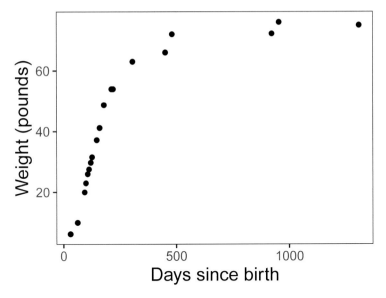

FIGURE 2.2 Measured weight of the dog Wilson over time.

Wow! The code to produce Figure 2.2 looks complicated. Let's break this down step by step:

- `ggplot(data = wilson) +` sets up the graphics structure and identifies the name of the data frame we are including.
- `geom_point(aes(x = days, y = weight))` defines the type of plot we are going to be making.
- `geom_point()` defines the type of plot geometry (or *geom*) we are using here - in this case, a point plot.
- `aes(x = days, y = weight)` maps the *aesthetics* of the plot. On the x axis is the `days` variable; on the y axis is the `weight` variable. You may also write this as `mapping = aes(x = days, y = weight)`.
- The statement beginning with `labs(x=...)` defines the labels on the x and y axes.

I know this seems like a lot of code to make a visualization, but this structure is actually used for some more advanced data visualization. Think of the `+` structure at the end of each line as the connector between `ggplot` and the plot `geom`. Trust me - learning how to make informative plots can be a useful skill!

2.5.2 Making a line plot

Using the same `wilson` data, later on we will discover that the function $W = f(D) = \dfrac{70}{1 + e^{2.46-0.017D}}$. represents these data. In order to make a graph of this function we need to first build a data frame (Figure 2.3):

```r
# Choose spacing that is "smooth enough"
days <- seq(from = 0, to = 1500, by = 1)
weight <- 70 / (1 + exp(2.46 - 0.017 * days))

wilson_model <- tibble(
  days = days,
  weight = weight
)

ggplot(data = wilson_model) +
  geom_line(aes(x = days, y = weight)) +
  labs(
    x = "Days since birth",
    y = "Weight (pounds)"
  )
```

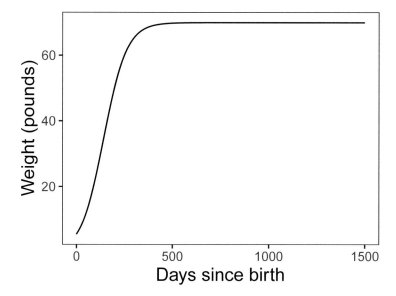

FIGURE 2.3 Logistic model equation to describe the weight of the dog Wilson over time.

Notice that once we have the data frame set up, the structure is very similar to the scatter plot - but this time we are using geom_line() rather than geom_point.

2.5.3 Changing options

Curious about using a different color in your plot or a thicker line? That is fairly easy to do. For example if we wanted to make either our points or line a different color, we adjust the ggplot to the following code (not evaluated here, but try it out on your own):

```
ggplot(data = wilson) +
  geom_point(aes(x = days, y = weight), color = "red", size = 2)
labs(
  x = "Days since birth",
  y = "Weight (pounds)"
)
```

Notice how the command color='red' was applied *outside* of the aes - which means it gets mapped to each of the points in the data frame. size=2 refers to the size (in millimeters) of the points. I've linked more options about the colors and sizes you can use here:

- **Named colors in R:** gallery of R colors.[13] Scroll down to "Picking one color in R" - you can see the list of options!
- **More colors:** colors in ggplot.[14]. More information about working with colors.
- **Using hexadecimal colors:** hexadecimal colors.[15] (You specify these by the code so "#FF3300" is a red color.)
- **Changing sizes of lines and points:** modifying a ggplot.[16]

2.5.4 Combining scatter and line plots.

Combining the data (Figure 2.2) with the model (Figure 2.3) in the same plot can be done by combining the geom_point with the geom_line, as shown in the following code (try it out on your own):

```
ggplot(data = wilson) +
  geom_point(aes(x = days, y = weight), color = "red") +
  geom_line(data = wilson_model, aes(x = days, y = weight)) +
  labs(
    x = "Days since birth",
    y = "Weight (pounds)"
  )
```

Notice in the above code a subtle difference when I added in the dataset wilson_model with geom_line: you need to name the data bringing in a new

---

[13]https://www.r-graph-gallery.com/42-colors-names.html
[14]http://www.cookbook-r.com/Graphs/Colors_(ggplot2)/
[15]http://www.cookbook-r.com/Graphs/Colors_(ggplot2)/#hexadecimal-color-code-chart
[16]https://ggplot2.tidyverse.org/articles/ggplot2-specs.html

data frame to a plot geom. While it may be useful to have a plot legend[17], for this textbook the context will be apparent without having a legend.

2.6   Defining functions

We will study lots of other built-in functions for this course, but you may also be wondering how you define your own function (let's say $y = x^3$). We need the following construct for our code:

```
function_name <- function(inputs) {

  # Code

  return(outputs)
}
```

Here function_name serves as what you call the function, inputs are what you need in order to run the function, and outputs are what gets returned. So if we are doing $y = x^3$ then we will call that function cubic:

```
cubic <- function(x) {
  y <- x^3
  return(y)
}
```

So now if we want to evaluate $y(2) = 2^3$ at the console we type cubic(2). Neat! The following code will make a plot of the function $y = x^3$ using cubic (try this out on your own):

```
x <- seq(from = 0, to = 2, by = 0.05)
y <- cubic(x)

my_data <- tibble(x = x, y = y)

ggplot(data = my_data) +
  geom_line(aes(x = x, y = y)) +
  labs(
    x = "x",
    y = "y"
  )
```

---

[17]http://www.cookbook-r.com/Graphs/Legends_(ggplot2)/

2.6.1 Functions with multiple inputs

Sometimes you may want to define a function with different input parameters, so for example the function $y = x^3 + c$. To define that, we can modify the function to have input variables:

```
cubic_revised <- function(x, c) {
  y <- x^3 + c
  return(y)
}
```

To create and plot several examples of the function `cubic` for different values of c is shown in the following code and Figure 2.4.

```
x <- seq(from = 0, to = 2, by = 0.05)


my_data_revised <- tibble(
  x = x,
  c_zero = cubic_revised(x, 0),
  c_pos1 = cubic_revised(x, 1),
  c_pos2 = cubic_revised(x, 2),
  c_neg1 = cubic_revised(x, -1)
)


ggplot(data = my_data_revised) +
  geom_line(aes(x = x, y = c_zero)) +
  geom_line(aes(x = x, y = c_pos1)) +
  geom_line(aes(x = x, y = c_pos2)) +
  geom_line(aes(x = x, y = c_neg1)) +
  labs(
    x = "x",
    y = "y"
  )
```

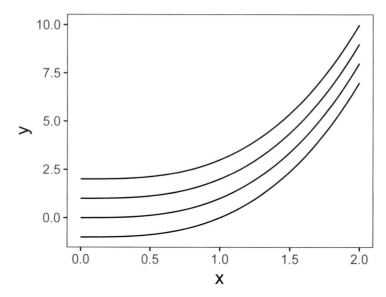

FIGURE 2.4 Plot of several cubic functions $y = x^3 + c$ when $c = -1, 0, 1, 2$.

Notice how I defined multiple columns of the data frame `my_data_revised` in the `tibble` command, and then used mutiple `geom_line` commands to plot the data. Since we had combined the different values of `c` in a single data frame we didn't need to define the `data` with each instance of `geom_line`.

2.7   Concluding thoughts

This is not meant to be a self-contained chapter in R but rather one so that you can quickly compute to code. Curious to learn more? Thankfully there are several good resources. Here are few of my favorites that I turn to:

- **R Graphics.**[18] This is a go-to resource for making graphics. (I also use Google a lot too.)
- **The Pirates Guide to R.**[19] This book promises to build your R knowledge from the ground up.
- **R for Reproducible Scientific Analysis.**[20] This set of guided tutorials can help you build your programming skills in R.

---

[18]`http://www.cookbook-r.com/`
[19]`https://bookdown.org/ndphillips/YaRrr/`
[20]`http://swcarpentry.github.io/r-novice-gapminder/`

- **R for Data Science**.[21] This is a useful book to take your R knowledge to the next level.

The best piece of advice: DON'T PANIC! Patience and persistence are your friend. Reach out for help, and recognize that like with any new endeavor, practice makes progress.

2.8 Exercises

Exercise 2.1. Create a folder on your computer and a project file where you will store all your R work.

Exercise 2.2. Install the packages devtools, tidyverse to your R installation. Once that is done, then install the package demodelr.

Exercise 2.3. What are the variables listed in the dataset phosphorous in the demodelr library? (Hint: try the command ?phosphorous.)

Exercise 2.4. Make a scatterplot (geom_point()) of the dataset phosphorous in the demodelr library. Be sure to label the axes with descriptive titles.

Exercise 2.5. Change Figure 2.3 so the line is blue and the size is 4 mm.

Exercise 2.6. Change the color of the points in Figure 2.2 to either a hexadecimal color or a named color of your choice.

Exercise 2.7. For this exercise you will do some plotting:

a. Define a sequence (call this sequence x) that ranges between -12 to 12 with spacing of .05.
b. Also define the variable y such that $y = \sin(x)$.
c. Make a scatter plot to graph $y = \sin(x)$. Set the points to be red.
d. Make a line plot to graph $y = \sin(x)$. Label the x-axis with your favorite book title. Label the y-axis with your favorite food to eat.

Exercise 2.8. An equation that relates a consumer's nutrient content (denoted as y) to the nutrient content of food (denoted as x) is given by: $y = cx^{1/\theta}$, where $\theta \geq 1$ and $c > 0$ are both constants. Let's just assume that $c = 1$ and the $0 \leq x \leq 1$.

a. Construct a function called nutrient that will make a sequence of y values for an input x and theta (θ).

---

[21] https://r4ds.had.co.nz/

b. Use your `nutrient` function to create a line plot (`geom_line()`) for five different values of $\theta > 1$, appropriately labeling all axes.

Exercise 2.9. The dataset `phosphorous` in the `demodelr` library contains measurements of the phosphorous content of *Daphnia* and its primary food source algae.

Researchers believe that *Daphnia* has strict homeostatic regulation of the phosphorous contained in algae, and want to determine the value of θ in the equation $y = y = cx^{1/\theta}$. They have already determined that the value of $c = 1.737$.

a. Complete Exercise 2.4. Be sure to label the axes correctly.
b. Use your function `nutrient` from Exercise 2.8 to make an initial guess for theta (θ) that is consistent with the data. You can evaluate your guess by plotting (with `geom_line()`) against the data.
c. Use guess and check to refine the value of θ that seems to work best.
d. Report your value of θ.

Exercise 2.10. For this exercise you will investigate some built-in functions. Remember you can learn more about a function by typing `?FUNCTION`, where `FUNCTION` is the name.

a. Explain (using your own words) what the function `runif(1,100,1000)` does.
b. Explain (using your own words) what the function `ceiling()` does, showing an example of its use.

Exercise 2.11. The Ebola outbreak in Africa in 2014 severely affected the country of Sierra Leone. A model for the number of Ebola infections I is given by the following equation:
$$I(t) = \frac{K \cdot I_0}{I_0 + (K - I_0)\exp(-rt)},$$

where $K = 13580$, $I_0 = 251$ and $r = 0.0227$. The variable t is in days. Use `geom_line()` to visualize this curve from $0 \le t \le 700$.

Exercise 2.12. Consider the following piecewise function:

$$y = \begin{cases} x^2 & \text{for } 0 \le x < 1, \\ 2 - x & \text{for } 1 \le x \le 2 \end{cases} \tag{2.1}$$

a. Define a function in R that computes y for $0 \le x \le 2$.
b. Use `geom_line()` to generate a graph of $y(x)$ over the interval $0 \le x \le 2$.

Exercise 2.13. An insect's development rate r depends on temperature T (degrees Celsius) according to the following equation:

$$r = \begin{cases} 0.1 & \text{for } 17 \leq T < 27, \\ 0 & \text{otherwise.} \end{cases} \qquad (2.2)$$

a. Define a function in R that computes r for $0 \leq T \leq 30$.

b. Use geom_line() to generate a graph of $r(T)$ over the interval $0 \leq T \leq 30$.

3

Modeling with Rates of Change

Chapter 1 provided examples for modeling with rates of change, and Chapter 2 introduced the computational and visualization software R and RStudio, and how we can translate equations with rates of change to understand phenomena. The focus for this chapter will be on taking a contextual description and starting to develop differential equation models for them.

Oftentimes when we construct differential equations from a contextual description we bring our own understanding and knowledge to this situation. How *you* may write down the differential equation may be different from someone else - *do not worry!* This is the fun part of modeling: models can be considered testable hypotheses that can be refined when confronted with data. Let's get started

3.1 Competing plant species and equilibrium solutions

Consider the following context to develop a mathematical model:

A newly introduced plant species is introduced to a region. It competes with another established species for nutrients (and is a better competitor). However, the growth rate of the new species is proportional to the difference between the current number of established species and the number of new species. You may assume that the number of established species is a constant E.

For this problem we will start by naming our variables. Let N represent number of new species and E the number of established species. We will break this down accordingly:

- *"the growth rate of the new species"* describes the rate of change, or derivative, expressed as $\dfrac{dN}{dt}$.

- *"is proportional to the difference between the current number of established species and the number of new species"* means $\alpha \cdot (E - N)$, where α is the proportionality constant. Including this parameter helps to avoid assuming we have a 1:1 correspondence between the growth rate of the new species and the population difference.

- *"and is a better competitor"* helps to explain why the term is $\alpha \cdot (E - N)$ instead of $\alpha \cdot (N - E)$. We know that the newly established species will start out in much smaller numbers than N. But since it is a better competitor, we would expect its rate to increase initially. So $\dfrac{dN}{dt}$ should be *positive* rather than negative.

Taking all these assumptions together, Equation (3.1) shows the differential equation to model this context:

$$\frac{dN}{dt} = \alpha \cdot (E - N) \tag{3.1}$$

You may recognize that Equation (3.1) is similar to Equation (1.4) in Chapter 1 for the spread of Ebola. It is not surprising to have similar differential equations appear in different contexts. We will see throughout this book that it is more advantageous to learn techniques to analyze models qualitatively rather than memorize several different types of models and not see the connections between them.

An interesting solution to a differential equation is the *steady state* or *equilibrium solution*. Equilibrium solutions occur where the rates of change are zero. For Equation (3.1), this means that we are solving $\dfrac{dE}{dt} = \alpha \cdot (E - N) = 0$. Granted, the expression $\alpha \cdot (E - N)$ may look like alphabet soup, but it is helpful to remember that α and E are both parameters; the steady state occurs when the expression $E - N$ equals zero, or when $N = E$. We may consider the new species N to be established when it reaches the same population level as E. Identifying steady states in a model aids in understanding the behavior of any solutions for a differential equation. Chapters 5 and 6 dig deeper into steady states and their calculation.

3.2   The Law of Mass Action

Our next example focuses on how to generate a model that borrows concepts from modeling chemical reactions. For example let's say you have a substrate A

that reacts with enzyme B to form a product S. One common way to represent this process is with a reaction equation (Equation (3.2)):

$$A + B \rightarrow S \tag{3.2}$$

Figure 3.1 is a schematic diagram of Equation (3.2):

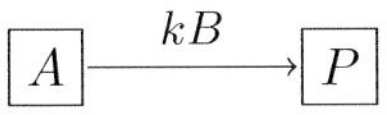

FIGURE 3.1 Schematic diagram of a substrate-enzyme reaction.

One key quantity is the rate of formation for the product P, which we express by Equation (3.3):

$$\frac{dP}{dt} = kAB, \tag{3.3}$$

where k is the proportionality constant or the rate constant associated with the reaction. Notice how we express the interaction between A and B as a product - if either the substrate A or enzyme B is not present (i.e. A or B equals zero), then product P is not formed. Equation (3.3) is an example of the law of mass action.

Modeling interactions (whether between susceptible and infected individuals, enzymes and substrates, or predators and prey) with the law of mass action is always a good first assumption to understand the system, which can be subsequently refined. For example, if we consider that the substrate might decay, we can revise Figure 3.1 to Figure 3.2:

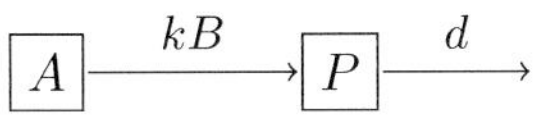

FIGURE 3.2 Revised schematic diagram of substrate-enzyme reaction with decay of the product P.

In this instance the rate of change of P would then include a term dP (Equation (3.4)):

$$\frac{dP}{dt} = kAB - dP \tag{3.4}$$

3.3   Coupled differential equations: lynx and hares

Another example is a *system of differential equations*. The context is between
the snowshoe hare and the Canadian lynx, shown in Figure 3.3. Figure 3.4 also
displays a timeseries of the two populations overlaid. Notice how in Figure
3.4 both populations show regular periodic fluctuations. One plausible reason
is that the lynx prey on the snowshoe hares, which causes the population to
initially decline. Once the snowshoe hare population declines, then there is less
food for the lynx to survive, so their population declines. The decline in the
lynx population causes the hare population to increase, and the cycle repeats.[1]

FIGURE 3.3 Examples of lynx and hare - aren't they beautiful?

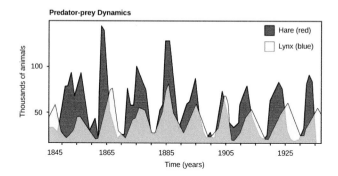

FIGURE 3.4 Timeseries of the combined lynx and hare populations. Notice
how the populations are coupled with each other.

In summary it is safe to say that the two populations are *coupled* to one another,
yielding a coupled system of equations. But in order to understand how they
are coupled together, first let's consider the two populations *separately*.

---

[1]There is a lot more nuance for reasons behind periodic fluctuations in these two
populations, which includes more complicated food web interactions and climate variation.
MacLulich (1937), Stenseth et al. (1997), and King and Schaffer (2001) are good places to
dig into the complexity of this fascinating biological system.   Image sources for Figure 3.3:
Kilby (2012) and Frank, Jacob W. (2021). Image source for Figure 3.4: OpenStax (2016)

To develop the mathematical model we will make some simplifying assumptions. The hares grow much more quickly than then lynx - in fact some hares have been known to reproduce several times a year. A reasonable assumption for large hare populations is that rate of change of the hares is proportional to the hare population. Based on this assumption Equation (3.5) describes the rate of change of the hare population, with H as the population of the hares:

$$\frac{dH}{dt} = rH \tag{3.5}$$

Since the growth rate r is positive, so then the rate of change (H') will be positive as well, and H will be increasing. A representative value for r is 0.5 year^{-1} (Mahaffy 2010; Brady and Butler 2021). You may be thinking that the units on r seem odd - (year^{-1}), but that unit on r makes the term rH dimensionally consistent to be a rate of change.

Let's consider the lynx now. An approach is to assume their population declines exponentially, or changes at the rate proportional to the current population. Let's consider L to be the lynx population, with the following differential equation (Equation (3.6)):

$$\frac{dL}{dt} = -dL \tag{3.6}$$

We assume the death rate d in Equation (3.6) is positive, leading to a negative rate of change for the Lynx population (and a decreasing value for L). A typical value of d is 0.9 yr^{-1} (Mahaffy 2010; Brady and Butler 2021).

The next part to consider is how the lynx and hare interact. Since the hares are prey for the lynx, when the lynx hunt, the hare population decreases. We can represent the process of hunting with the following adjustment to our hare equation:

$$\frac{dH}{dt} = rH - bHL \tag{3.7}$$

So the parameter b represents the hunting rate. Notice how we have the term HL for this interaction. This term injects a sense of realism: if the lynx are not present ($L = 0$), then the hare population can't decrease due to hunting. We model the *interaction* between the hares and the lynx with multiplication between the H and L. A typical value for b is .024 lynx^{-1} year^{-1}. It is okay if that unit seems a little odd to you - it should be! As before, if we multiply out the units on bHL we would get units of hares per year.

How does hunting affect the lynx population? One possibility is that it increases the lynx population:

$$\frac{dL}{dt} = bHL - dL \tag{3.8}$$

Notice the symmetry between the rate of change for the hares and the lynx equations. In many cases this makes sense - if you subtract a rate from one population, then that rate should be added to the receiving population. You could also argue that there is some efficiency loss in converting the hares to lynx - not all of the hare is converted into lynx biomass. In this situation we sometimes like to adjust the hunting term for the lynx equation with another parameter e, representing the efficiency that hares are converted into lynx:

$$\frac{dL}{dt} = e\,bHL - dL \qquad (3.9)$$

(sometimes people just make a new parameter $c = e\,b$, but for now we will just leave it as is and set $e = 0.2$). Equation (3.10) shows the coupled system of differential equations:

$$
\begin{aligned}
\frac{dH}{dt} &= rH - bHL \\
\frac{dL}{dt} &= e\,bHL - dL
\end{aligned}
\qquad (3.10)
$$

The schematic diagram representing these interactions is shown in Figure 3.5:

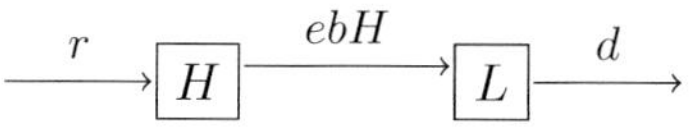

FIGURE 3.5 Schematic diagram Lynx-Hare system.

Equation (3.10) is a classical model in mathematical biology and differential equations - it is called the *predator-prey* model, also known as the *Lotka-Volterra model* (Lotka 1920, 1926; Volterra 1926).

3.4   Functional responses

In several examples we have seen a rate of change proportional to the current population, as, for example, the rate of growth of the hare population is rH. This is one example of what we would call a functional response[2]. Another type of functional response assumes that the rate reaches a limiting value proportional to the population size, so $\frac{dH}{dt} = \frac{rH}{1 + arH}$. This is an example of a **type II functional response**. Finally, the type II response has also been

---

[2]https://en.wikipedia.org/wiki/Functional_response

generalized (a **type III functional response**) $\dfrac{dH}{dt} = \dfrac{rH^2}{1 + arH^2}$. Figure 3.6 shows all three functional responses together:

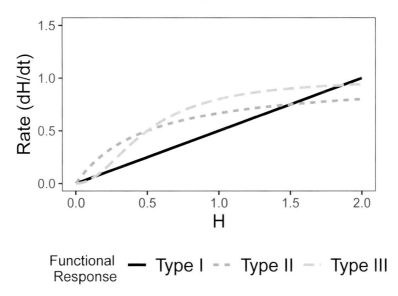

FIGURE 3.6 Comparison between examples of Type I - Type III functional responses. For a Type I functional response the rate grows proportional to population size H, whereas for Types II and III the rate reaches a saturating value.

Notice the limiting behavior in the Type II and Type III functional responses. These responses are commonly used in ecology and predator-prey dynamics and in problems of how animals search for food.

3.5 Exercises

Exercise 3.1. Consider the following types of functional responses:

$$\text{Type I:} \quad \frac{dP}{dt} = 0.1P$$
$$\text{Type II:} \quad \frac{dP}{dt} = \frac{0.1P}{1 + .03P} \tag{3.11}$$
$$\text{Type III:} \quad \frac{dP}{dt} = \frac{0.1P^2}{1 + .05P^2}$$

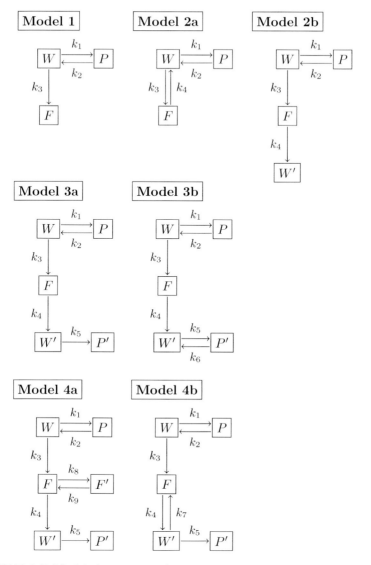

FIGURE 3.7 Modeled reaction schemes representing the potential effect of a pesticide on water quality.

Exercise 3.7. (Inspired by Burnham and Anderson (2002)) You are tasked with the job of investigating the effect of a pesticide on water quality, in terms of its effects on the health of the plants and fish in the ecosystem. Different models can be created that investigate the effect of the pesticide. Different types of reaction schemes for this system are shown in Figure 3.7, where F represents the amount of pesticide in the fish, W the amount of pesticide in the water, and S the amount of pesticide in the soil. The prime (e.g. F', W', and S' represent other bound forms of the respective state). In all seven different models can be derived. For each of the model schematics, apply the Law of Mass Action to write down a system of differential equations.

4

Euler's Method

Chapter 3 examined modeling with rates of change. Once a differential equation model is defined one possible next step is to determine the solution to the differential equation. While in some cases an exact solution can be found (Chapter 7), in many instances we will rely on numerical methods.

The focus of this chapter is on *approximation* of solutions to a differential equation via a numerical method. Typically a first numerical method you might learn is *Euler's method*, popularized in the movie Hidden Figures.[1] This chapter will develop Euler's method from tangent line equations or locally linear approximations from calculus. Let's get started!

4.1 The flu and locally linear approximation

Consider Equation (4.1), which is one way to model the rate of change of the flu through a population:

$$\frac{dI}{dt} = 3e^{-.025t} \qquad (4.1)$$

In Equation (4.1) the variable I represents the number of people infected at day t. One question we could address using Equation (4.1) is to predict the value of I after 1 day, assuming that $I(0) = 10$. To do that we will build a locally linear approximation at $t = 0$ and use the approximation to forecast and estimate $I(1)$.

In order to solve this problem, the formula for the locally linear approximation to $I(t)$ at $t = 0$ is $L(t) = I(0) + I'(0) \cdot (t - 0)$. Here, $I(0) = 10$ and $I'(0) = 3$ (found by evaluating Equation (4.1) at $t = 0$). Using $L(t) \approx I(t)$, the formula for the locally linear approximation is given by Equation (4.2). To define Equation (4.2) we used two pieces of information: the (given) value of the function at $t = 0$ and the estimate of the derivative from Equation (4.1).

---

[1] https://www.youtube.com/watch?v=v-pbGAts_Fg

$$L(t) = 10 + 3t \tag{4.2}$$

At $t = 1$ we can make a prediction with Equation (4.2) to estimate that there will be 13 people sick. To evaluate this approximation it is helpful to compare our prediction from $L(1)$ (Equation (4.2)) to the actual value from the solution to the differential equation given in Equation (4.3):

$$I(t) = 130 - 120e^{-.025t} \tag{4.3}$$

Table 4.1 compares the values of the linear approximation (Equation (4.2)) to Equation (4.3):

TABLE 4.1 Comparison of the exact solution $I(t)$ (Equation (4.3)) to the linear approximation $L(t)$ (Equation (4.2)) at $t = 0$ and $t = 1$.

t	Linear approximation $L(t)$	Exact solution $I(t)$
0	10	10
1	13	12.96

Table 4.1 shows that $L(1)$ is an *overestimate* compared to $I(1)$. Let's expand Equation (4.2) even more by constructing *another* linear approximation using the differential equation. We will denote this linear approximation as $L_1(t)$ to distinguish it from $L(t)$ from Equation (4.2). First we evaluate Equation (4.1), which yields $I'(1) = 2.92$. The formula for the linear approximation at $t = 1$ is $L_1(t) = I(1) + I'(1) \cdot (t - 1)$. Here we will use $I(1) = 13$, recognizing that this value is a pretty close estimate for the number infected (I) at $t = 1$. This assumption yields $L_1(t) = 13 + 2.92(t - 1)$.

We can continue to build out the solution in a similar manner to develop a locally linear approximation at $t = 2$, shown graphically in Figure 4.1. The approximation and the exact solution in Figure 4.1 appear very close to each other, suggesting that approximation using local linearization could work for other types of differential equations.

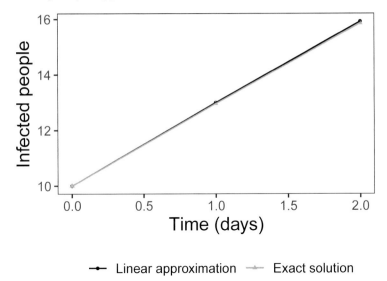

FIGURE 4.1 Approximation of a solution to Equation (4.1) using local linearity.

By eye, the approximate and exact solutions in Figure 4.1 appear indistinguishable from each other. Encouraged by these results, let's develop the approach with linear approximations even more.

4.2 A workflow for approximation

The previous chapter alludes to a possible workflow to numerically approximate a solution to a differential equation:

- Determine the locally linear approximation at a given point.
- Forecast out to another time value.
- Repeat the locally linear approximation.

The results of continuing this workflow (approximate → forecast → repeat) several times is shown in Table 4.2.

Comparison of the exact solution $I(t)$ (Equation (4.3)) to the linear approximation $L(t)$ (Equation (4.2)) at $t = 0$ and $t = 1$.

TABLE 4.2 Comparison of the exact solution $I(t)$ (Equation (4.3)) to forecasting with linear approximations at $t = 90$ and $t = 95$.

t	Approximate solution	Exact solution $I(t)$
90	118.4	117
95	119.9	118.6

Table 4.2 suggests that the accuracy of our solution decreases as time increases. A potential fix would be to approximate the solution not at every day, but every half day. The length of time that we forecast out our solution is called the step size, denoted as Δt. While approximating our solution every half day ($\Delta t = 0.5$) would require more computation (or more iterations) of the locally linear approximation, perhaps a smaller Δt would lead to more accurate solutions. Let's start out smaller with the first few timesteps (Table 4.3):

TABLE 4.3 Calculation of the solution $I(t)$ for Equation (4.1) using the linear approximations at each timestep with $\Delta t = 0.5$.

t	I	$\dfrac{dI}{dt}$	$\dfrac{dI}{dt} \cdot \Delta t$
0	10	3	1.5
0.5	$= 10 + 1.5 = 11.5$	2.96	1.48
1	$= 11.5 + 1.48 = 12.98$	2.92	1.46
1.5	$= 12.92 + 1.46 = 14.38$	2.88	1.44
2	$= 14.38 + 1.44 = 15.82$		

Notice how Table 4.3 organizes a way to compute the solution I with linear approximations. Each row is a "step" of the method, computing the solution based on our step size Δt. The third column computes the value of the derivative for a particular time (Equation (4.1)), and then the fourth column represents the forecasted change in the solution by the next timestep.[2]

This idea of *approximate, forecast, repeat* is at the heart of many numerical methods[3] that approximate solutions to differential equations. The particular method that we have developed here is called *Euler's method*. We display the results from additional steps in Figure 4.2. Based on the trend of the solution in Figure 4.2, it appears that the number of infections might start to level off at $I = 130$, which is the steady state value in Equation (4.3) when evaluating $\lim\limits_{t \to \infty} I(t)$.

---

[2] When you have a *rate of change* multiplied by a time increment this will give you an approximation of the net change in a function.

[3] https://en.wikipedia.org/wiki/Numerical_methods_for_ordinary_differential_equations

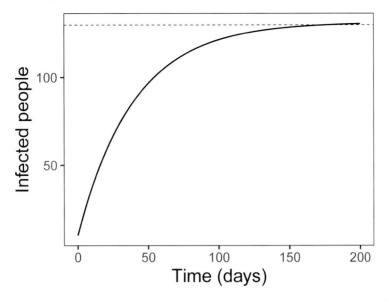

FIGURE 4.2 Longer-term approximation of a solution to Equation (4.1). Notice how the solution seems to level off to a steady state at $I = 130$ (dashed line).

4.3 Building an iterative method

Now that we have worked on an example, let's carefully formulate Euler's method with another example. In Chapter 1 we discussed the spread of Ebola through a population. Equation (4.5) is the differential equation for the logistic model (see Equation (1.8) and Figure 1.5), modeled with Equation (4.4):

$$\frac{dI}{dt} = 0.023I \cdot (13600 - I), \tag{4.4}$$

where the variable I represents the proportion of people that are infected. The *carrying capacity*, or the place where the solution levels off in Equation (4.4) is at $I = 13600$ (notice that when $I = 13600$, $\frac{dI}{dt} = 0$). Numerical methods such as Euler's method can become unstable for large values of the independent variable, because the rates are so large. To account for this, we will re-define Equation (4.4) with the variable $p = \frac{I}{13600}$, leading to the revised model:

$$\frac{dp}{dt} = 0.023p \cdot (1 - p), \tag{4.5}$$

where the variable p represents the proportion of infected, So $p = 1$ means that 13600 people are infected. Once we have our solution $p(t)$, we can just multiply that by $N = 13600$ to return back to the total infected.

In Equation (4.5) we define the function $f(p) = 0.023p \cdot (1 - p)$. In order to numerically approximate the solution, we will need to recall some concepts from calculus. This first step is that we will approximate the rate of change $\dfrac{dp}{dt}$ with a difference quotient (Equation (4.6)):

$$\frac{dp}{dt} = \lim_{\Delta t \to 0} \frac{p(t + \Delta t) - p(t)}{\Delta t} \tag{4.6}$$

When the quantity Δt in Equation (4.6) is small (for example $\Delta t = 1$ day), this difference quotient provides a reasonable way to organize the problem:

$$\frac{p(t + \Delta t) - p(t)}{\Delta t} = 0.023p \cdot (1 - p)$$
$$p(t + \Delta t) - p(t) = 0.023p \cdot (1 - p) \cdot \Delta t \tag{4.7}$$
$$p(t + \Delta t) = p(t) + 0.023p \cdot (1 - p) \cdot \Delta t$$

The last expression $(p(t + \Delta t) = p(t) + 0.023p \cdot (1 - p) \cdot \Delta t)$ defines an iterative system, easily computed with a spreadsheet program, or with a `for` loop in R:

```
# Define your timestep and time vector
deltaT <- 1
t <- seq(0, 600, by = deltaT)

# Define the number of steps we take. This is equal to 10 / dt (why?)
N <- length(t)

# Define current solution state:
p_approx <- 250/13600

# Define a vector for your solution:the derivative equation
for (i in 2:N) { # We start this at 2 because the first value is 10
  dpdt <- .023 * p_approx[i - 1] * (1 - p_approx[i - 1])
  p_approx[i] <- p_approx[i - 1] + dpdt * deltaT
}


# Define your data for the solution into a tibble:
solution_data <- tibble(
  time = t,
  prop_infected = p_approx
)
```

```
# Plot your solution:
ggplot(data = solution_data) +
  geom_line(aes(x = time, y = prop_infected)) +
labs(
  x = "Time (days)",
  y = "Proportion infected"
)
```

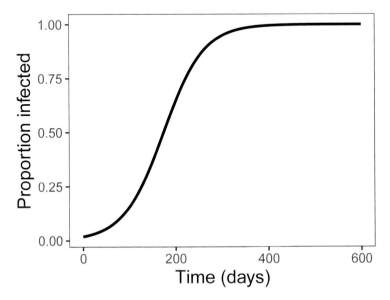

FIGURE 4.3 Results from applying an iterative method to solve Equation (4.5).

Let's break the code down that produced Figure 4.3 step by step:[4]

- `deltaT <- 0.1` and `t <- seq(0,2,by=deltaT)` define the timesteps (Δt) and the output time vector `t`.
- The statement `N <- length(t)` defines how many steps we take.
- `p_approx<- 250/13600` defines the *proportion* of the population initially infected (assuming that $I(0) = 250$). We will use this as the starting point to the solution vector.
- The `for` loop goes through this system - first computing the value of $\frac{dp}{dt}$ and then forecasing out the next timestep $p(t + \Delta t) = f(p) \cdot \Delta t$. We iteratively build the vector `p_approx`, adding another element at each timestep.
- The remaining code plots the data frame, like we learned in Chapter 2.

---

[4] You may notice that when you run the code to produce Figure 4.3 on your own it may not look like the output shown here. I've customized how the plots are displayed in this book using the pacakge ggthemes.

4.3.1 Euler's method in demodelr

To generate Figure 4.3 we created the solution directly in R - but you don't want to copy and paste the code. The demodeler package has a function called euler that does the same process to generate the output solution:[5] Try running the following code and plotting your solution:

```r
# Define the rate equation:
infection_eq <- c(dpdt ~ .023 * p * (1 - p))

# Define the initial condition (as a named vector):
prop_init <- c(p = 250/13600)

# Define deltaT and the time steps:
deltaT <- 1
n_steps <- 600

# Compute the solution via Euler's method:
out_solution <- euler(system_eq = infection_eq,
                      initial_condition = prop_init,
                      deltaT = deltaT,
                      n_steps = n_steps
                      )
```

Once the vector out_solution is created, it has variables t and p, which can then be plotted with a ggplot statement. Let's talk through the steps of this code as well:

- The line infection_eq <- c(dpdt ~ .023 * p * (13600-i)) represents the differential equation, written in formula notation. So $\dfrac{dp}{dt} \to$ dpdt and $f(p) \to$ .023 * p * (1-p), with the variable p.
- The initial condition $p(0) = 250/13600 = .018$ is written as a **named vector:** prop_init <- c(p=250/13600). Make sure the name of the variable is consistent with your differential equation.
- As before we need to identify Δt (deltaT) and the number of steps N (n_steps). When we generated the solution in Figure 4.3, in the for loop we defined the ending point at $t = 2$ so the number of steps (N) was 20.[6]

The command euler then computes the solution applying Euler's method, returning a data frame so we can plot the results. Note the columns of the data frame are the variables t and i that have been named in our equations.

---

[5] Don't forget to load up the demodelr library in your code at the top of your R code.
[6] In general if we know Δt and the time we wish to end computing (t_{end}, then $N = t_{end}/\Delta t$.

4.3.2 Euler's method applied to systems

Now that we have some experience with Euler's method, let's see how we can apply the function `euler` to a system of differential equations. Here is a sample code that shows the dynamics for the lynx-hare equations, as studied in Chapter 3:

$$\frac{dH}{dt} = rH - bHL$$
$$\frac{dL}{dt} = ebHL - dL \tag{4.8}$$

The variables H and L are already in thousands of animals, so we don't need to rescale anything like we did with Equation (4.5). We are going to use Euler's method to solve this differential equation, using the code below:

```
# Define the rate equation:
lynx_hare_eq <- c(
  dHdt ~ r * H - b * H * L,
  dLdt ~ e * b * H * L - d * L
)

# Define the parameters (as a named vector):
lynx_hare_params <- c(r = 2, b = 0.5, e = 0.1, d = 1)

# Define the initial condition (as a named vector):
lynx_hare_init <- c(H = 1, L = 3)

# Define deltaT and the number of time steps:
deltaT <- 0.05
n_steps <- 200

# Compute the solution via Euler's method:
out_solution <- euler(system_eq = lynx_hare_eq,
                      parameters = lynx_hare_params,
                      initial_condition = lynx_hare_init,
                      deltaT = deltaT,
                      n_steps = n_steps
                      )

# Make a plot of the solution,
# using different colors for lynx or hares:
ggplot(data = out_solution) +
  geom_line(aes(x = t, y = H), color = "red") +
  geom_line(aes(x = t, y = L), color = "blue",linetype='dashed') +
  labs(
```

```
    x = "Time",
    y = "Lynx (red) or Hares (blue/dashed)"
)
```

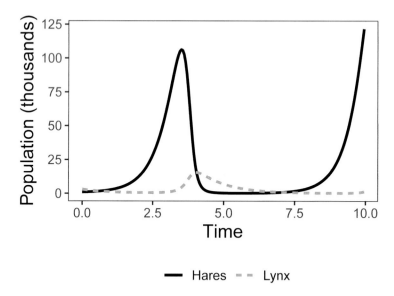

FIGURE 4.4 Euler's method solution for Lynx-Hare system (Equation 4.8).

This example is structured similarly when we used Euler's method to solve a single variable differential equation, with some key changes (that are easy to adapt):

- The variable `lynx_hare_eq` is now a vector, with each entry one of the rate equations.
- We need to identify both variables in their initial condition.
- Most importantly, Equation (4.8) has parameters, which we define as a named vector `lynx_hare_params <- c(r = 2, b = 0.5, e = 0.1, d = 1)` that we pass through to the command `euler` with the option `parameters`. If your equation does not have any parameters you do not need to worry about specifying this input.
- We plot both solutions together at the end, or you can make two separate plots. Remember that you can choose the color in your plot. I included the additional option `linetype=dashed` for the hares population for ease of viewing.

4.4 Euler's method and beyond

Sometimes when working with Euler's method you encounter a differential equation that produces some nonsensible results. For example, consider a model that represents infection with quarantine (see Exercise 1.10 in Chapter 1):

$$\frac{dS}{dt} = -kSI$$
$$\frac{dI}{dt} = -kSI - \beta I$$

(4.9)

In Equation (4.9), susceptibles become sick by encountering an infected person, but infected people are removed from the population at a rate β. The model in Figure 4.5 illustrates the results when this model is implemented using `euler`:

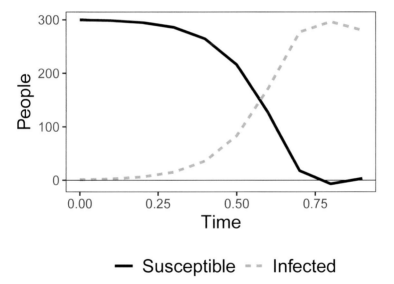

FIGURE 4.5 Surprising results when using Euler's method to solve Equation (4.9). Notice how some values for I are negative.

You may notice in Figure 4.5 the solution for S falls below $S = 0$ around $t = 0.75$.[7] Negative values for S are concerning because we know there can't be negative people!

---

[7] Notice in the code I've added a line for the horizontal axis with geom_hline.

At a given timestep, Euler's method constructs a locally linear approximation and forecasts the solution forward to the next timestep. Using Figure 4.5, at $t = 0.75$ the value for $S \approx 1$ and the value for $I \approx 280$. If we let $k = 0.05$ and $\beta = 0.2$, this means that $\dfrac{dS}{dt} = -14$ and $\dfrac{dI}{dt} = -42$. At this point, the values of S and I are both decreasing. In turn, the forecast value for S at $t = 0.75$ is $S = 1 - 14 \cdot 0.1 = -0.4$. Mathematically, Euler's method is working correctly, but we know realistically that neither S nor I can be negative.

While Euler's method is useful, it does quite poorly in cases where the solution is changing rapidly, such as described above. A way to circumvent this is to adjust the value of Δt to be smaller, which comes at the expense of more computational time. A second way is to use a *higher order solver* than `euler`, and one such method is called the Runge-Kutta method[8]. (You study these methods when you take a course in numerical analysis. How we implement the Runge-Kutta method is to replace the command `euler` with `rk4`:

```
# Define the rate equation:
quarantine_eq <- c(
  dSdt ~ -k * S * I,
  dIdt ~ k * S * I - beta * I
)


# Define the parameters (as a named vector):
quarantine_parameters <- c(k = .05, beta = .2)


# Define the initial condition (as a named vector):
quarantine_init <- c(S = 300, I = 1)


# Define deltaT and the number of time steps:
deltaT <- .1 # timestep length
n_steps <- 10 # must be a number greater than 1


# Compute the solution via Runge-Kutta method:
out_solution <- rk4(system_eq = quarantine_eq,
                    parameters = quarantine_parameters,
                    initial_condition = quarantine_init,
                    deltaT = deltaT,
                    n_steps = n_steps
                    )


# Make a plot of the solution:
ggplot(data = out_solution) +
  geom_line(aes(x = t, y = S), color = "red") +
  geom_line(aes(x = t, y = I), color = "blue",linetype="dashed") +
```

---

[8]https://en.wikipedia.org/wiki/Runge%E2%80%93Kutta_methods

```
geom_hline(yintercept=0,size=0.25) +
labs(
  x = "Time",
  y = "Susceptible (red) or Infected (blue/dashed)"
)
```

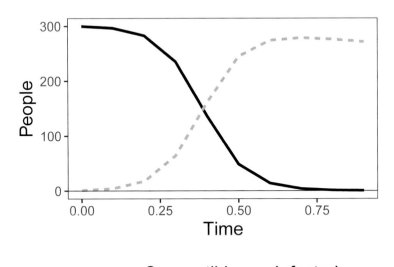

FIGURE 4.6 Runge-Kutta solution for Equation 4.9. Notice how the solution curve for the variable S does not fall below zero as it does in Figure 4.5.

Another benefit to the `rk4` method is the numerical error when computing the solution. The numerical error is quantified as the difference between the actual solution and the numerical solution. Euler's method has an error on the order of the stepsize Δt, whereas the Runge-Kutta method has an error of $(\Delta t)^4$. For this example, $\Delta t = 0.1$, whereas for the Runge-Kutta method the numerical error is on the order of 0.0001 $((\Delta t)^4 = .0001)$ - noticeably different! We can improve Euler's method by taking a smaller timestep - BUT that means we need a larger number of steps N - which may take more computational time (see Exercise 4.15). Does this discussion of numerical error sounds familiar? In calculus you may have examined the numerical error when using Riemann sums (left, right, trapezoid, midpoint sums) to approximate the area underneath a curve. While the context is different, Riemann sums and numerical differential equation solvers are closely related.

In summary, the most general form of a differential equation is:

$$\frac{d\vec{y}}{dt} = f(\vec{y}, \vec{\alpha}, t), \tag{4.10}$$

where in Equation (4.10) $\vec{y}$ is the vector of state variables you want to solve for, and $\vec{\alpha}$ is your vector of parameters. At a given initial condition, Euler's method applies locally linear approximations to forecast the solution forward Δt time units (Equation (4.11)):

$$\vec{y}_{n+1} = \vec{y}_n + f(\vec{y}_n, \vec{\alpha}, t_n) \cdot \Delta t \qquad (4.11)$$

Both the Euler or Runge-Kutta methods define a workflow (approximate $\rightarrow$ forecast $\rightarrow$ repeat) to generate a numerical solution to a system of differential equations. The process of defining a workflow is a powerful technique that we will revisit several times throughout this textbook, so stay tuned!

4.5 Exercises

Exercise 4.1. Verify that $I(t) = 130 - 120e^{-0.025t}$ is a solution to the differential equation

$$\frac{dI}{dt} = 130 - 0.025I$$

with $I(0) = 10$.

Exercise 4.2. Apply the rk4 solver with $\Delta t = 1$ with $N = 600$ to the initial value problem $\dfrac{dp}{dt} = 0.023p \cdot (13600 - p)$, $p(0) = 250/13600$. Compare your solution to Figure 4.3. What differences do you observe? Which solution method (euler or rk4) is better (and why)?

Exercise 4.3. In the model presented by Equation (4.9), is $S + I$ constant? *Hint: add* $\dfrac{dS}{dt}$ *and* $\dfrac{dI}{dt}$.

Exercise 4.4. The following exercise will help you explore the relationships between stepsize, ending points, and number of steps needed. You may assume that we will start at $t = 0$ in all parts.

a. If we wish to do a Euler's method solution with step size 1 second and ending at $T = 5$ seconds, how many steps will we take?
b. If we wish to do a Euler's method solution with step size 0.5 seconds and ending at $T = 5$ seconds, how many steps will we take?
c. If we wish to do a Euler's method solution with step size 0.1 seconds and ending at $T = 5$ seconds, how many steps will we take?
d. If we wish to do a Euler's method solution with step size Δt and go to ending value of T, what is an expression that relates the number steps N as a function of Δt and T?

Exercise 4.5. To get a rough approximation between error and step size, let's say for a particular differential equation that we are starting at $t = 0$ and going to $t = 2$, with $\Delta t = 0.2$ with 10 steps. We know that the Runge-Kutta error will be on the order of $(\Delta t)^4 = 0.0016$. If we want to use Euler's method with the same order of error, we could say $\Delta t = .0016$. For that case, how many steps will we need to take?

Exercise 4.6. For each of the following differential equations, apply Euler's method to generate a numerical solution to the differential equation and plot your solution. The stepsize (Δt) and number of iterations (N) are listed.

a. Differential equation: $\dfrac{dS}{dt} = 3 - S$. Set $\Delta t = 0.1$, $N = 50$. Initial conditions: $S(0) = 0.5$, $S(0) = 5$.

b. Differential equation: $\dfrac{dS}{dt} = \dfrac{1}{1 - S}$. Set $\Delta t = 0.01$, $N = 30$. Initial conditions: $S(0) = 0.5$, $S(0) = 2$.

c. Differential equation: $\dfrac{dS}{dt} = 0.8 \cdot S \cdot (10 - S)$. Set $\Delta t = 0.1$, $N = 50$. Initial conditions: $S(0) = 3$, $S(0) = 10$.

Exercise 4.7. For each of the following differential equations, apply the Runge-Kutta method to generate a numerical solution to the differential equation and plot your solution. The stepsize (Δt) and number of iterations (N) are listed. Contrast your answers with Exercise 4.6.

a. Differential equation: $\dfrac{dS}{dt} = 3 - S$. Set $\Delta t = 0.1$, $N = 50$. Initial conditions: $S(0) = 0.5$, $S(0) = 5$.

b. Differential equation: $\dfrac{dS}{dt} = \dfrac{1}{1 - S}$. Set $\Delta t = 0.01$, $N = 30$. Initial conditions: $S(0) = 0.5$, $S(0) = 2$.

c. Differential equation: $\dfrac{dS}{dt} = 0.8 \cdot S \cdot (10 - S)$. Set $\Delta t = 0.1$, $N = 50$. Initial conditions: $S(0) = 3$, $S(0) = 10$.

Exercise 4.8. Complete the following steps:

a. Apply the code `euler` to generate a numerical solution to the differential equation:

- Differential equation: $\dfrac{dS}{dt} = r \cdot S \cdot (K - S)$.
- Set $r = 1.2$ and $K = 3$.
- Set $\Delta t = 0.1$, $N = 50$.
- Initial conditions (three different ones): $S(0) = 1$, $S(0) = 3$, $S(0) = 5$.

b. Plot your Euler's method solutions with the three initial conditions on the same plot. What do you notice when you do plot them together?

c. Make a hypothesis regarding the long term behavior of this system. Then plot a few more solution curves to verify your guess.

Exercise 4.9. Complete the following steps:

a. Apply the code `euler` to generate a numerical solution to the differential equation:

- Differential equation: $\dfrac{dS}{dt} = K - S$.
- Set $K = 2$.
- Set $\Delta t = 0.1$, $N = 50$.
- Initial conditions (three different ones): $S(0) = 0$, $S(0) = 2$, $S(0) = 5$.

b. Plot your Euler's method solutions with the three initial conditions on the same plot. What do you notice when you do plot them together?

c. Make a hypothesis regarding the long term behavior of this system. Then plot a few more solution curves to verify your guess.

Exercise 4.10. This exercise uses the following differential equation:

$$\frac{dS}{dt} = 0.8 \cdot S \cdot (10 - S) \tag{4.12}$$

a. Apply Euler's method with $S(0) = 15$, $\Delta t = 0.1$, $N = 10$.

b. When you examine your solution, what is incorrect about the Euler's method solution based on your qualitative knowledge of the underlying dynamics?

c. Now calculate Euler's method for the same differential equation for the following conditions: $S(0) = 15$, $\Delta t = 0.01$, $N = 100$. What has changed in your solution?

Exercise 4.11. Apply Euler's method to the differential equation $\dfrac{dS}{dt} = \dfrac{1}{1 - S}$ with the following conditions:

- $S(0) = 1.5$, $\Delta t = 0.1$, $N = 10$
- $S(0) = 1.5$, $\Delta t = 0.01$, $N = 100$.

Between these two solutions, what has changed? Do you think it is numerically possible to calculate a reasonable solution for Euler's method near $S = 1$? (*note: this differential equation is an example of finite time blow up*)

Exercise 4.12. One way to model the growth rate of hares is with $f(H) = \dfrac{rH}{1 + kH}$, where r and k are parameters. This is in constrast to exponential growth, which assumes $f(H) = rH$.

a. First evaluate $\lim\limits_{H \to \infty} rH$.

b. Then $\lim\limits_{H \to \infty} \dfrac{rH}{1 + kH}$.

c. Compare your two answers. Discuss how the growth rate $f(H) = \dfrac{rH}{1 + kH}$ seems to be a more realistic model.

Exercise 4.13. In the lynx-hare example we can also consider an alternative system where the growth of the hare is not exponential:

$$\frac{dH}{dt} = \frac{2H}{1 + kH} - 0.5HL$$
$$\frac{dL}{dt} = 0.05HL - L$$

(4.13)

Set the number of timesteps to be 2000, $\delta t = 0.1$, with initial condition $H = 1$ and $L = 3$. Apply Euler's method to numerically solve this system of equations when $k = 0.1$ and $k = 1$ and plot your simulation results.

Exercise 4.14. Consider the differential equation $\dfrac{dS}{dt} = \dfrac{1}{1 - S}$. Notice that at $S = 1$ the rate $\dfrac{dS}{dt}$ is not defined.

a. If you applied Euler's method solution with initial condition $S(0) = 0.9$, what would the values of S approach as time increases?
b. If you applied Euler's method solution with initial condition $S(0) = 1.1$, what would the values of S approach as time increases?
c. Explain how you could come to the same conclusion as the previous two problems if you graphed $f(S) = \dfrac{1}{1 - S}$.

Exercise 4.15. Building on Exercise 4.5, let's say for a particular differential equation we have N steps from $0 \le t \le b$. An error of ϵ is desired.

a. What is the ratio $\dfrac{N_E}{N_{RK4}}$, where N_{RK4} represents the number of steps needed for the Runge-Kutta method, and N_E the number of steps for Euler's method?
b. Make a plot of the ratio $\dfrac{N_E}{N_{RK4}}$ for $0 \le \epsilon \le 1$. How many more steps does Euler's method need to do to achieve the same level of error, compared to the Runge-Kutta method?

5

Phase Lines and Equilibrium Solutions

Chapter 4 explored numerical techniques to solve initial value problems. This chapter takes a step back to examine the general family of solutions to a differential equation. Will the family of solutions converge in the long run (as $t \to \infty$) to a constant value? Are these specific solutions that always remain constant (or independent of time)? Answering questions such as these address the *qualitative* behavior for a single differential equation.[1] Let's get started!

5.1 Equilibrium solutions

Chapter 3 introduced the concept of an equilibrium solution, or where the rate of change for a differential equation is zero. We can determine equilibrium solutions for a single-variable differential equation by setting the left hand side of $\dfrac{dy}{dt} = f(y)$ equal to zero and solving for y (or whatever dependent variable describes the problem).

Example 5.1. What are the equilibrium solutions to $\dfrac{dy}{dt} = -y$?

Solution. For this example we know that when the rate of change is zero, this means that $\dfrac{dy}{dt} = 0$, or when $0 = -y$. So $y = 0$ is the equilibrium solution.

The general solution to the differential equation $\dfrac{dy}{dt} = -y$ is $y(t) = Ce^{-t}$, where C is an arbitrary constant. (We will explore techniques to determine this in Chapter 7.) Figure 5.1 plots different initial conditions, with the equilibrium solution shown as a horizontal line:

---

[1] Qualitative behavior for coupled systems of differential equations is addressed in Chapter 6.

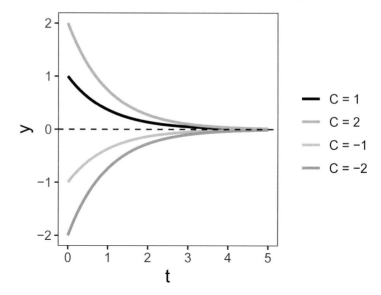

FIGURE 5.1 Solution curves to $y' = -y$ for different initial conditions (values of C).

Notice that in Figure 5.1 as t increases, all solutions approach the equilibrium solution $y = 0$, regardless if the initial condition is positive or negative. This observation is also confirmed by evaluating the limit $\lim_{t \to \infty} Ce^{-t}$, which is 0.

Example 5.2. Determine equilibrium solutions to

$$\frac{dN}{dt} = N \cdot (1 - N) \tag{5.1}$$

Solution. In this case the equilibrium solutions for Equation (5.1) occur when $N \cdot (1 - N) = 0$, or when $N = 0$ or $N = 1$.

The general solution to Equation (5.1) is

$$N(t) = \frac{C}{C + (1 - C)e^{-t}}. \tag{5.2}$$

Figure 5.2 displays several different solution curves for Equation (5.2).

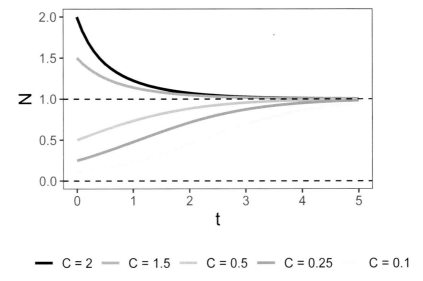

FIGURE 5.2 Solution curves for Equation (5.1) with different initial conditions (values of C).

In Figure 5.2 notice how all the solutions tend towards $N = 1$, but even solutions that start close to $N = 0$ seem to move away from the $N = 0$ equilibrium solution. Solutions in Figure 5.2 exhibit the idea of the *stability* of an equilibrium solution, which we discuss next.

5.2 Phase lines for differential equations

The **stability** of an equilibrium solution describes the long-term behavior of the family of solutions. Solutions can converge to the equilibrium solution in the long run, or they may not. More formally stated:

An equilibrium solution y_* to a differential equation $\dfrac{dy}{dt} = f(y)$ is considered *stable* when for a given solution $\lim_{t \to \infty} y(t) = y_*$.

You may note that the definition of stability relies on determining the solution

$y(t)$. However we can circumvent determining this solution by using ideas from calculus and the rate of change:

- If $\dfrac{dy}{dt} < 0$, the solution $y(t)$ is decreasing.
- If $\dfrac{dy}{dt} > 0$, the solution $y(t)$ is increasing.

So to classify stability of an equilibrium solution we can investigate the behavior of the differential equation *around* the equilibrium solutions.

Let's apply this logic to our differential equation $\dfrac{dy}{dt} = -y$ from Example 5.1.

When $y = 3$, $\dfrac{dy}{dt} = -3 < 0$, so we say the function is *decreasing* to $y = 0$.

When $y = -2$, $\dfrac{dy}{dt} = -(-2) = 2 > 0$, so we say the function is *increasing* to $y = 0$. This can be represented neatly in the *phase line diagram* for Figure 5.3.[2]

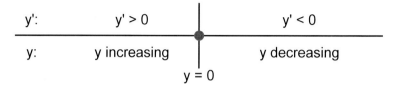

FIGURE 5.3 Phase line for the differential equation $y' = -y$.

Because the solution is *increasing* to $y = 0$ when $y < 0$, and *decreasing* to $y = 0$ when $y > 0$, we say that the equilibrium solution for the differential equation $y' = -y$ is **stable**, which is also confirmed by the solutions plotted in Figure 5.1.

Now let's generalize the example $y' = -y$ to classify the stability of the equilibrium solutions to $\dfrac{dy}{dt} = ry$, where r is a parameter. Fortunately the equilibrium solution is still $y = 0$. We will need to consider three different cases for the stability depending on the value of r ($r > 0$, $r < 0$, and $r = 0$):

- When $r < 0$, the phase line will be similar to Figure 5.3.
- When $r > 0$ the phase line will be as shown in Figure 5.4. We say in this case that the equilibrium solution is *unstable*, as all solutions flow away from the equilibrium. Several different solutions are shown in Figure 5.5 .
- When $r = 0$ we have the differential equation $\dfrac{dy}{dt} = 0$, which has $y = C$ as

---

[2]Sometimes arrows are used in the phase line to signify if the solutions are increasing or decreasing. I will stick to the convention presented in Figure 5.3 because it illustrates connections between the differential equation and the solution.

a general solution. For this special case the equilibrium solution is neither stable or unstable[3].

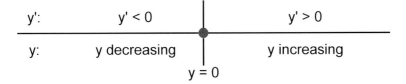

FIGURE 5.4 Phase line for the differential equation $y' = ry$, with $r > 0$.

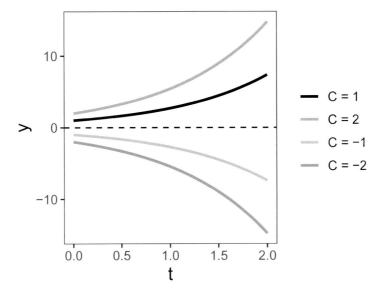

FIGURE 5.5 Solution curves for the differential equation $y' = ry$, with $r > 0$ for different initial conditions (values of C).

Based on the above discussion, let's return back to the differential equation $\frac{dN}{dt} = N \cdot (1 - N)$ from Example 5.2. We evaluate the stability of the equilibrium solutions $N = 0$ and $N = 1$ in Table 5.1.

---

[3] Arguably when $r = 0$ the resulting differential equation $y' = 0$ is different than $y' = ry$; something peculiar is going on here - which is discussed more in Chapter 20.

TABLE 5.1 Evaluation of the stability of the equilibrium solutions for Equation (5.1).

Test point	Sign of N'	Tendency of solution
$N = -1$	Negative	Decreasing
$N = 0$	Zero	Equilibrium solution
$N = 0.5$	Positive	Increasing
$N = 1$	Zero	Equilibrium solution
$N = 2$	Negative	Decreasing

Notice how the selected test points in the first column of Table 5.1 were selected to the *left* or the *right* of the equilibrium solutions. The phase line diagram of Figure 5.6 also presents the same information as in Table 5.1, but in contrast to Figures 5.3 and 5.4 we need to include *two* equilibrium solutions. Phase line diagrams should include all the computed equilibrium solutions.

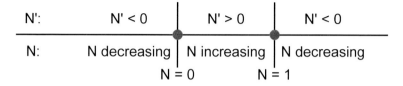

FIGURE 5.6 Phase line diagram for Equation (5.1).

Table 5.1 and Figure 5.6 confirm that solutions move *away* from the equilibrium solution $N = 0$ and move *towards* the equilibrium solution $N = 1$. These results suggest that the equilibrium solution at $N = 1$ is *stable* and the equilibrium solution at $N = 0$ is *unstable*. Therefore, one way to define an equilibrium solution y_* as unstable is when $\lim_{t \to -\infty} y(t) = y_*$.

5.3   A stability test for equilibrium solutions

Notice how when constructing the phase line diagram we relied on the behavior of solutions *around* the equilibrium solution to classify the stability. As an alternative we can also use the point at the equilibrium solution itself.

Consider the general differential equation $\dfrac{dy}{dt} = f(y)$ with an equilibrium solution at y_*. Next we apply local linearization to construct a locally linear approximation to $L(y)$ to $f(y)$ at $y = y_*$ (Equation (5.3)):

$$L(y) = f(y_*) + f'(y_*) \cdot (y - y_*) \tag{5.3}$$

There are two follow-on steps to simplify Equation (5.3). First, because we have an equilibrium solution, $f(y_*) = 0$. Second, Equation (5.3) can be written with a new variable P, defined by variable $P = y - y_*$. With these two steps Equation (5.3) translates to Equation (5.4):

$$\frac{dP}{dt} = f'(y_*) \cdot P \tag{5.4}$$

Does Equation (5.4) look familiar? It should! This equation is similar to the example where we classified the stability of $\frac{dy}{dt} = ry$ (notice that $f'(y_*)$ is a number). Using this information, a test to classify the stability of an equilibrium solution is the following:

Local linearization stability test for equilibrium solutions: For a differential equation $\frac{dy}{dt} = f(y)$ with equilibrium solution y_*, we can classify the stability of the equilibrium solution through the following:

- If $f'(y_*) > 0$ at an equilibrium solution, the equilibrium solution $y = y_*$ will be *unstable*.
- If $f'(y_*) < 0$ at an equilibrium solution, the equilibrium solution $y = y_*$ will be *stable*.
- If $f'(y_*) = 0$, we cannot conclude anything about the stability of $y = y_*$.

Let's return back to the differential equation $\frac{dN}{dt} = N \cdot (1 - N)$ from Example 5.2 and apply the local linearization stability test, $f'(N) = 1 - 2N$. Since $f'(0) = 1$, which is greater than 0, the equilibrium solution $N = 0$ is unstable. Likewise, if $f'(1) = -1$, the equilibrium solution $N = 1$ is stable.

Applying the local linearization test may be easier to quickly determine stability of an equilibrium solution. Guess what? This test also is a simplified form of determining stability of equilibrium solutions for systems of differential equations. We will explore this more in Chapter 19.

5.4 Exercises

Exercise 5.1. For the following differential equations, (1) determine any equilibrium solutions, and (2) classify the stability of the equilibrium solutions by applying the local linearization test.

a. $\dfrac{dS}{dt} = 0.3 \cdot (10 - S)$

b. $\dfrac{dP}{dt} = P \cdot (P - 1)(P - 2)$

Exercise 5.2. Using your results from Exercise 5.1, construct a phase line for each of the differential equations and classify the stability of the equilibrium solutions.

Exercise 5.3. A population grows according to the equation $\dfrac{dP}{dt} = \dfrac{P}{1 + 2P} - 0.2P$.

a. Determine the equilibrium solutions for this differential equation.
b. Classify the stability of the equilibrium solutions using the local linearization stability test.

Exercise 5.4. (Inspired by Logan and Wolesensky (2009)) A cell with radius r assimilates nutrients at a rate proportional to its surface area, but uses nutrients proportional to its volume, according to the following differential equation:

$$\frac{dr}{dt} = k_A 4\pi r^2 - k_V \frac{4}{3}\pi r^3, \tag{5.5}$$

where k_A and k_V are positive constants.

a. Determine the equilibrium solutions for this differential equation.
b. Construct a phase line for this differential equation to classify the stability of the equilibrium solutions.
c. Classify the stability of the equilibrium solutions using the local linearization stability test. Are your conclusions the same from the previous part?

Exercise 5.5. (Inspired by Thornley and Johnson (1990)) The Chanter equation of growth is the following, where W is the weight of an object:

$$\frac{dW}{dt} = W(3 - W)e^{-Dt} \tag{5.6}$$

Use this differential equation to answer the following questions.

a. What happens to the rate of growth ($\dfrac{dW}{dt}$) as t grows large?
b. What are the equilibrium solutions to this model? Are they stable or unstable?
c. Notice how the equilbrium solutions are the same as those for the logistic model. Based on your previous work, why would this model be a more realistic model of growth than the logistic model $\dfrac{dW}{dt} = W(3 - W)$?

Exercise 5.6. Red blood cells are formed from stem cells in the bone marrow. The red blood cell density r satisfies an equation of the form

$$\frac{dr}{dt} = \frac{br}{1+r^n} - cr, \tag{5.7}$$

where $n > 1$ and $b > 1$ and $c > 0$. Find all the equilibrium solutions r_* to this differential equation. *Hint:* can you factor an r from your equation first?

Exercise 5.7. (Inspired by Hugo van den Berg (2011)) Organisms that live in a saline environment biochemically maintain the amount of salt in their blood stream. An equation that represents the level of S in the blood is the following:

$$\frac{dS}{dt} = I + p \cdot (W - S), \tag{5.8}$$

where the parameter I represents the active uptake of salt, p is the permeability of the skin, and W is the salinity in the water.

a. First set $I = 0$. Determine the equilibrium solutions for this differential equation. Express your answer S_* in terms of the parameters p and W.
b. Next consider $I > 0$. Determine the equilibrium solutions for this differential equation. Express your answer S_* in terms of the parameters p, W, and I. Why should your new equilbrium solution be greater than the equilibrium solution from the previous problem?
c. Classify the stability of both equilibrium solutions in both cases using the local linearization stability test.

Exercise 5.8. (Inspired by Logan and Wolesensky (2009)) The immigration rate of bird species (species per time) from a mainland to an offshore island is $I_m \cdot (1 - S/P)$, where I_m is the maximum immigration rate, P is the size of the source pool of species on the mainland, and S is the number of species already occupying the island. Further, the extinction rate is $E \cdot S/P$, where E is the maximum extinction rate. The growth rate of the number of species on the island is the immigration rate minus the extinction rate, given by the following differential equation:

$$\frac{dS}{dt} = I_m \left(1 - \frac{S}{P}\right) - \frac{ES}{P} \tag{5.9}$$

a. Determine the equilibrium solutions S_* for this differential equation. Expression your answer in terms of I_M, P, and E.
b. Classify the stability of the equilibrium solutions using the local linearization stability test.

Exercise 5.9. A colony of bacteria growing in a nutrient-rich medium depletes the nutrient as they grow. As a result, the nutrient concentration $x(t)$ is steadily decreasing. The equation describing this decrease is the following:

$$\frac{dx}{dt} = -\mu \frac{x \cdot (\xi - x)}{\kappa + x}, \tag{5.10}$$

where μ, κ, and ξ are all parameters greater than zero.

a. Determine the equilibrium solutions x_* for this differential equation.
b. Construct a phase line for this differential equation and classify the stability of the equilibrium solutions.

Exercise 5.10. Can a solution curve cross an equilibrium solution of a differential equation?

6

Coupled Systems of Equations

Chapter 5 focused on qualitative analysis of a single differential equation using phase lines and slope fields. This chapter extends this idea further to systems of differential equations, where the natural extension of a phase line is a *phase plane*. Here is the good news: many of the techniques are similar to the ones introduced in Chapter 5, so let's get started!

6.1 Flu with quarantine and equilibrium solutions

In Exercise 1.10 in Chapter 1 we developed the following model for the flu as a coupled system of equations shown in Equation (6.1):

$$\frac{dS}{dt} = -kSI$$
$$\frac{dI}{dt} = kSI - \beta I \qquad (6.1)$$
$$\frac{dR}{dt} = \beta I,$$

where S represents susceptible people, I infected people, and R recovered people. Another way to represent this context is with the schematic shown in Figure 6.1:

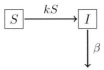

FIGURE 6.1 Schematic of the flu model with quarantine.

While Equation (6.1) is a system of three differential equations, notice that the variable R is not present on the right hand sides of each equation. As a result, the variable R is decoupled from this system of equations, so we can just focus on the rates of change for S and I (Equation (6.2)):

$$\frac{dS}{dt} = -kSI$$

$$\frac{dI}{dt} = kSI - \beta I \qquad (6.2)$$

With Equation (6.2) we will solve for equilibrium solutions (similar to what we did in Chapters 3 and 5), which we focus on next.

6.2   Nullclines

The process to determine equilibrium solutions for a system of differential equations starts with computing the *nullclines* for each rate in the system of equations. The nullclines are solutions in the plane where one of the rates is zero, so for example either $\frac{dS}{dt}$ or $\frac{dI}{dt}$ is zero. For coupled systems of equations, the equilibrium solutions are where the rates $\frac{dS}{dt}$ and $\frac{dI}{dt}$ in Equation (6.2) are *both* zero, found through algebraically solving the system of equations in Equation (6.3):

$$0 = -kSI$$

$$0 = kSI - \beta I \qquad (6.3)$$

Let's examine the first equation ($0 = -kSI$). Since all the terms are expressed as a product, then nullclines for S occur when either $S = 0$ or $I = 0$.

In a similar manner, the nullclines for I occur when $0 = kSI - \beta I$. For this expression we can factor out an I, yielding $0 = I \cdot (kS - \beta)$. Because the last equation is factored as a product, nullclines for I are either $I = 0$ or by solving $kS - \beta$ for S to yield $S = \frac{\beta}{k}$.

Nullclines are not equilibrium solutions by themselves - it is the *intersection* of two different nullclines that determines equilibrium solutions. Figure 6.2 shows the nullclines in the $S - I$ plane (since we have two equations), with S on the horizontal axis and I on the vertical axis. In Figure 6.2 we have also assumed that $\beta = 1$ and $k = 1$. The $S - I$ plane shown in Figure 6.2 is the beginning of the construction of the phase plane for Equation (6.2) and also to determine the equilibrium solutions.

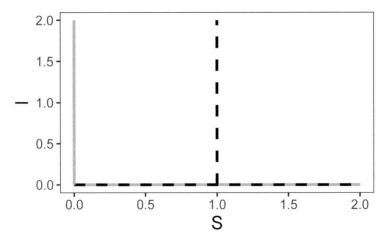

Nullclines: — S' = 0 — I' = 0

FIGURE 6.2 Nullclines for Equation (6.2). To generate the plot we assumed $\beta = 1$ and $k = 1$.

A key thing to note is that where two different nullclines cross is an *equilibrium solution* to the system of equations (**both** $\dfrac{dS}{dt}$ and $\dfrac{dI}{dt}$ are zero at this point). Examining Figure 6.2 three possibilities appear:

1. There is an equilibrium solution at $S = 0$ and $I = 0$ (otherwise known as the origin). This equilibrium solution makes biological sense: if there is nobody susceptible or infected there are no flu cases (everyone is perfectly healthy - yay!) .

2. The entire horizontal axis is an equilibrium solution because $I = 0$, which makes both $\dfrac{dS}{dt}$ and $\dfrac{dI}{dt}$ zero. There is a practical interpretation of this nullcline - whenever $I = 0$, meaning there are no infected people around, infection cannot occur.

3. There is also a third possibility where the vertical line at $S = 1$ crosses the horizontal axis ($S = 1$, $I = 0$), but that also falls under the second equilibrium solution.[1]

Now that we have identified our nullclines and equilibrium solutions, we will add additional context with the *flow* of the solution.

---

[1]For the general system (Equation (6.2)), the equilibrium solution would that be $S = \dfrac{\beta}{k}$ and $I = 0$.

TABLE 6.1 Values of $\dfrac{dS}{dt}$ (as 'dSdt') and $\dfrac{dI}{dt}$ (as 'dIdt') for Equation 6.2.

S	0	1	2	0	1	2	0	1	2
I	0	0	0	1	1	1	2	2	2
dSdt	0	0	0	0	-1	-2	0	-2	-4
dIdt	0	0	0	-1	0	1	-2	0	2

6.3   Phase planes

Next we can add more context to the Figure 6.2 by evaluating different values of S and I into our system of equations and plotting the *phase plane*. How we plot the phase plane is similar to the method in Chapter 5. We will test points around an equilibrium solution to determine if the solution is increasing or decreasing in S or I independently.

Table 6.1 evaluates the derivatives $\dfrac{dS}{dt}$ and $\dfrac{dI}{dt}$ in (6.2) for different values of S and I.

Notice in Table 6.1 the different values of $\dfrac{dS}{dt}$ and $\dfrac{dI}{dt}$ in Equation (6.2) at each of the given S and I values. We can plot each of the coordinate pairs of $\left(\dfrac{dS}{dt}, \dfrac{dI}{dt}\right)$ as a vector (arrows) in the (S, I) plane. To do so, associate $\dfrac{dS}{dt}$ with left-right motion, so positive values of $\dfrac{dS}{dt}$ mean the vector points to the right. Likewise, we associate $\dfrac{dI}{dt}$ with up-down motion, so positive values $\dfrac{dI}{dt}$ mean the vector points up.

Defining the directions of the vectors in this way is also consistent when Equation (6.2) is evaluated at the nullcline solutions. At the point $(S, I) = (1, 1)$, we have an arrow that points directly to the west because $\dfrac{dS}{dt} < 0$ and $\dfrac{dI}{dt} = 0$. Continuing on in this manner, by sequentially sampling points in the (S, I) plane we get a vector field plot (Figure 6.3), superimposed with the nullclines.

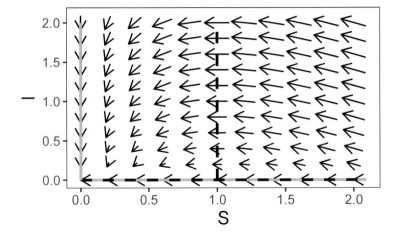

FIGURE 6.3 Phase plane for Equation (6.2), with $\beta = 1$ and $k = 1$.

6.3.1 Motion around the nullclines

We can also extend the motion around the nullclines to investigate the stability of an equilbrium solution. With a one-dimensional differential equation we used a number line to quantify values where the solution is increasing / decreasing. The problem with several differential equations is that the notion of "increasing" or "decreasing" becomes difficult to understand - there is an additional degree of freedom! Simply put, in a plane you can move left/right *or* up/down. The benefit for having nullclines is that they **isolate** the motion in one direction. When $\dfrac{dS}{dt} = 0$ the only allowed motion is up and down; when $\dfrac{dI}{dt} = 0$ the only allowed motion is left and right.

In general for a two-dimensional system:

- When a horizontal axis variable has a nullcline, the only allowed motion is up/down.
- When a vertical axis variable has a nullcline, the only motion is up/down.

Applying this knowledge to Equation (6.2), if we choose points where $I' = 0$ then we know that the only motion is to the left and the right because S can still change along that curve. If we choose points where $S' = 0$ then we know that the only motion is to the up/down because I can still change along that curve.

6.3.2   Stability of an equilbrium solution

Figure 6.3 qualitatively tells us about the stability of an equilibrium point.
One of the equilibrium solutions is at the origin $(S, I) = (0, 0)$. As before we
want to investigate if the equilibrium solution is stable or unstable. As you
can see the arrows appear to be pointing into and towards the equilibrium
solution. So we would classify this equilbrium solution as *stable*.

6.4   Generating a phase plane in R

Let's take what we learned from the case study of the flu model with quarantine
to qualitatively analyze a system of differential equations:

- We determine nullclines by setting the derivatives equal to zero.
- Equilibrium solutions occur where nullclines for the two different equations
  intersect.
- The arrows in the phase plane help us characterize the stability of the
  equilibrium solution.

The demodelr package has some basic functionality to generate a phase plane.
Consider the following system of differential equations (Equation (6.4)):

$$\frac{dx}{dt} = x - y$$
$$\frac{dy}{dt} = x + y$$
$$\tag{6.4}$$

In order to generate a phase plane diagram for Equation (6.4) we need to
define functions for x' and y', which I will annotate as dx and dy respectively.
We are going to collect these equations in one vector called system_eq, using
the tilde ($\sim$) as a replacement for the equals sign:

```
system_eq <- c(
  dx ~ x - y,
  dy ~ x + y
)
```

Then what we do is apply the command phaseplane, which will generate a
vector field over a domain:

```
phaseplane(system_eq,
  x_var = "x",
  y_var = "y"
)
```

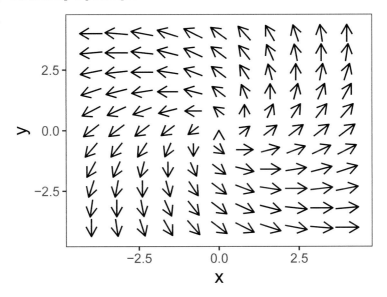

FIGURE 6.4 Phase plane for Equation (6.4).

Let's discuss how the phaseplane function works, first with the required inputs:

- You will need a system of differential equations (which we defined as system_eq).
- Next you need to define which variable belongs on the horizontal axis (x_var = 'x') or the vertical axis (y_var = 'y'). In Exercise 6.3 you will explore what happens if these get mixed up.

There are some additional options for phaseplane:

- The option eq_soln = TRUE will determine if there are any equilibrium solutions to be found and report them to the console. This option does not provide a definitive answer, but at least it tells you where to look. You can always confirm if a point is an equilibrium solution by evaluating the differential equation.
- You can adjust the windows that are plotted with the options x_window and y_window. Both of these need to be defined as a vector (e.g. x_window = c(-0.1,0.1). The default window size is [−4, 4] for both axes.
- There is an option parameters that allows you to pass any parameters to the phase plane. Later chapters will introduce systems where you can modify the parameters - we won't worry about that now.

6.5   Slope fields

For a one-dimensional differential equation, we call the phase plane a *slope field*. For a given differential equation $y' = f(t, y)$, at each point in the $t - y$ plane the differential equation is evaluated, showing the direction of the tangent line at that particular point. The phaseplane function can also plot slope fields. Let's take a look at an example first and then discuss how that it works.

Example 6.1. A colony of bacteria growing in a nutrient-rich medium depletes the nutrient as they grow. As a result, the nutrient concentration $x(t)$ is steadily decreasing. Determine the slope field for the following differential equation:

$$\frac{dx}{dt} = -0.7 \cdot \frac{x \cdot (3 - x)}{1 + x} \tag{6.5}$$

The R code shown below will generate the slope field for Equation (6.5) (shown in Figure 6.5):

```
# Define the windows where we make the plots
t_window <- c(0, 3)
x_window <- c(0, 5)

# Define the differential equation
system_eq <- c(
  dt ~ 1,
  dx ~ -0.7 * x * (3 - x) / (1 + x)
)

phaseplane(system_eq,
  x_var = "t",
  y_var = "x",
  x_window = t_window,
  y_window = x_window
) +
  theme_bw() +
  theme(
    legend.position = "bottom",
    legend.text = element_text(size = 14),
    axis.title.x = element_text(size = 14),
    axis.text.x = element_text(size = 10),
    axis.text.y = element_text(size = 10),
    axis.title.y = element_text(size = 14)
  ) +
  scale_color_colorblind()
```

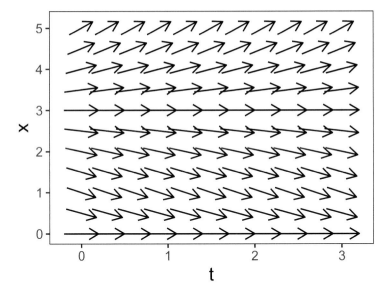

FIGURE 6.5 Slope field for Equation (6.5).

A few notes about the code that generated Figure 6.5:

- The variable on the horizontal axis (x_var) is t, and on the vertical axis (y_var) is x. Confusing, I know.
- The viewing window for the axis is also defined accordingly.
- Notice how the variable system_eq also contains the additional equation dt = 1. What we are doing is re-writing Equation (6.5) by introducing a new variable s (Equation (6.6)):

$$\frac{dt}{ds} = 1$$
$$\frac{dx}{ds} = -0.7 \cdot \frac{x \cdot (3-x)}{1+x}$$

(6.6)

The differential equation $\frac{dt}{ds} = 1$ has a solution $s = t$, so really Equation (6.6) is a (slightly more) complicated way to express Equation (6.5). Hacky? Perhaps. However re-writing Equation (6.5) was a quick and handy workaround to re-use code.

This chapter introduced a lot of useful R code to aid in visualization. The good news is that we will explore additional analyses of systems of differential equations starting with Chapter 15 - there is so much more to learn. Onward!

6.6   Exercises

Exercise 6.1. Determine equilibrium solutions for Equation (6.4).

Exercise 6.2. Generate a phase plane for Equation (6.4), but set the option eq_soln = TRUE. Did phaseplane detect the equilibrium solution you found in Exercise 6.1? If not, repeat the phase plane, but set x_window = c(-0.1,0.1) and y_window = c(-0.1,0.1) and repeat.

Exercise 6.3. Generate a phase plane for Equation (6.4), but this time set x_var = y and y_var = x (swap which variable is which). Notice that the incorrect phase plane is produced. What is the corresponding differential equation that is visualized by this phase plane?

Exercise 6.4. This problem considers the following system of differential equations:

$$\frac{dx}{dt} = y$$
$$\frac{dy}{dt} = -x \tag{6.7}$$

a. Determine the equations of the nullclines and equilibrium solution of this system of differential equations.
b. Modify the function phaseplane to generate a phase plane of this system.
c. For each point along a nullcline, determine the resulting motion (up-down or left-right).
d. Based on the work you generated, determine if the equilibrium solution is *stable*, *unstable*, or *inconclusive*.
e. Verify that the functions $x(t) = \sin(t)$ and $y = \cos(t)$ is one solution to this system of differential equations.

Exercise 6.5. Consider the following system of differential equations:

$$\frac{dx}{dt} = y$$
$$\frac{dy}{dt} = 3x^2 - 1 \tag{6.8}$$

a. Determine the equations of the nullclines and equilibrium solutions for this system of differential equations.
b. For each point along a nullcline, determine the resulting motion (up-down or left-right).
c. Modify the function phaseplane to generate a phase plane of this system. Adjust the windows for x and y to be between -1 and 1.

d. Make a hypothesis to classify if the equilibrium point is *stable* or *unstable*.

Exercise 6.6. (Inspired by Thornley and Johnson (1990)) A plant grows proportional to its current length L. Assume this proportionality constant is μ, whose rate also decreases proportional to its current value. The system of equations that models this plant growth is the following:

$$\frac{dL}{dt} = \mu L$$
$$\frac{d\mu}{dt} = -0.1\mu$$
(6.9)

a. Explain why $L = 0$ and $\mu = 0$ is an equilibrium solution to this differential equation.
b. Modify the function phaseplane to generate a phase plane of this system. Use the window $-0.1 \leq L \leq 0.1$ and $-0.1 \leq \mu \leq 0.1$. (For this problem negative values of L and μ are not sensible, but it aids in visualizing the equilibrium solution.)
c. Is the origin a stable equilibrium solution?

Exercise 6.7. (Inspired by Logan and Wolesensky (2009)) Red blood cells are formed from stem cells in the bone marrow. The red blood cell density r satisfies an equation of the form

$$\frac{dr}{dt} = \frac{0.2r}{1 + r^2} - 0.1r$$
(6.10)

a. What are the equilibrium solutions for this differential equation?
b. Modify the function phaseplane to generate a phase line for this differential equation for $0 \leq t \leq 5$ and $0 \leq r \leq 5$.
c. Based on the phase line, are the equilibrium solutions stable or unstable?

Exercise 6.8. (Inspired by Hugo van den Berg (2011)) Organisms that live in a saline environment biochemically maintain the amount of salt in their blood stream. An equation that represents the level of S in the blood is the following:

$$\frac{dS}{dt} = 1 + 0.3 \cdot (3 - S)$$

a. What are the equilibrium solutions for this differential equation?
b. Modify the function phaseplane to generate a phase line for this differential equation for $0 \leq t \leq 10$ and $0 \leq S \leq 10$.
c. Based on the phase line, are the equilibrium solutions stable or unstable?

Exercise 6.9. Consider the differential equation $\frac{dx}{dt} = -3x$. Here you will examine creating a two-dimensional system of equations by re-parameterizing $s = t$.

a. Define the variable $t = s$. For this case, what is $\dfrac{dt}{ds}$?

b. When $x = f(t(s))$ (x is a composition between t and s), one way to express the chain rule is $\dfrac{dx}{ds} = \dfrac{dx}{dt} \cdot \dfrac{dt}{ds}$. Use this fact to explain why $\dfrac{dx}{ds} = -3x$.

c. Finally use your previous work to determine the system of equations for $\dfrac{dx}{ds}$ and $\dfrac{dt}{ds}$.

Exercise 6.10. (Inspired by Hugo van den Berg (2011)) The core body temperature (T) of a mammal is coupled to the heat production (scaled by heat capacity Q) with the following system of differential equations:

$$\frac{dT}{dt} = Q + 0.5 \cdot (20 - T)$$
$$\frac{dQ}{dt} = 0.1 \cdot (38 - T) \tag{6.11}$$

a. Determine the equations of the nullclines and equilibrium solution of this system of differential equations.

b. For each point along a nullcline, determine the resulting motion (up-down or left-right). You may assume that both $T > 0$ and $Q > 0$.

c. Make a hypothesis to classify if the equilibrium solution is *stable* or *unstable*.

Exercise 6.11. Consider the following system of differential equations for the lynx-hare model (Equation (3.10) from Chapter 3):

$$\frac{dH}{dt} = rH - bHL$$
$$\frac{dL}{dt} = ebHL - dL \tag{6.12}$$

a. Determine the equilibrium solutions for this system of differential equations.

b. Determine equations for the nullclines, expressed as L as a function of H. There should be two nullclines for each rate.

Exercise 6.12. (Inspired by Hugo van den Berg (2011)) A chemostat is a tank used to study microbes and ecology, where microbes grow under controlled conditions. Think of this like a large tank with nutrient-rich water that is continuously cycled. For example, differential equations that describe the microbial biomass W and the nutrient concentration C (in the culture) are the following:

$$\frac{dW}{dt} = \mu W - F\frac{W}{V}$$
$$\frac{dC}{dt} = D \cdot (C_R - C) - S\mu\frac{W}{V}, \tag{6.13}$$

where we have the following parameters: μ is the per capita reproduction rate, F is the flow rate, V is the volume of the culture solution, D is the dilution rate, C_R is the concentration of nutrients entering the culture, and S is a stoichiometric conversion of nutrients to biomass.

a. Write the equations of the nullclines for this differential equation.
b. Determine the equilibrium solutions for this system of differential equations.
c. Generate a phase plane for this differential equation with the values $\mu = 1$, $D = 1$, $C_R = 2$, $S = 1$, and $V = 1$.
d. Classify the stability of the equilbrium solutions.

Exercise 6.13. A classical paper *Experimental Studies on the Struggle for Existence: I. Mixed Population of Two Species of Yeast* by Gause (1932) examined two different species of yeast growing in competition with each other. The differential equations given for two species in competition are:

$$\frac{dy_1}{dt} = -b_1 y_1 \frac{(K_1 - (y_1 + \alpha y_2))}{K_1}$$
$$\frac{dy_2}{dt} = -b_2 y_2 \frac{(K_2 - (y_2 + \beta y_1))}{K_2}, \tag{6.14}$$

where y_1 and y_2 are the two species of yeast with the parameters b_1, b_2, K_1, K_2, α, β describing the characteristics of the yeast species.

a. Determine the equilibrium solutions for this differential equation. Express your answer in terms of the parameters b_1, b_2, K_1, K_2, α, β.
b. Gause (1932) computed the following values of the parameters: $b_1 = 0.21827$, $b_2 = 0.06069$, $K_1 = 13.0$, $K_2 = 5.8$, $\alpha = 3.15$, $\beta = 0.439$. Using these values and your results from part a, what would be the predicted values for the equilibrium solutions? Is there anything odd about the values for these equilibrium solutions?
c. Use the function rk4 to solve this system of differential equations numerically and plot your solutions. Use initial conditions of $y_1(0) = .375$ and $y_2(0) = .291$, with $\Delta t = 1$ and $N = 600$.

7

Exact Solutions to Differential Equations

Chapters 4, 5, and 6 studied numerical and qualitative tools to analyze differential equations. Phase planes and slope fields helped to determine the long-term stability of an equilibrium solution. Beyond these approaches, it is also helpful to know the *exact* solution to a differential equation. In this chapter we will study three techniques to determine exact solutions to differential equations, making connections to some tools that you know from calculus. We briefly illustrate the methods but also have lots of exercises for you to practice these techniques, with problems in and out of a given context. Let's get started!

7.1   Verify a solution

The first approach is direct verification also known as the guess and check method. Consider the differential equation shown in Equation (7.1):

$$\frac{dS}{dt} = 0.7S \tag{7.1}$$

Direct verification starts with a candidate solution and then checks to see if the candidate solution is consistent with the differential equation. If we have an initial value problem (e.g. $S(0) = 4$), then the candidate solution is also consistent with the initial condition.

For example, let's verify if the function $\tilde{S}(t) = 5e^{0.7t}$ is a solution to Equation (7.1). We do this by differentiating $\tilde{S}(t)$, which, using our knowledge of calculus, is $0.7 \cdot 5e^{0.7t}$. Finally, by rearrangement, $\dfrac{d\tilde{S}}{dt} = 0.7 \cdot 5e^{0.7t} = 0.7e^{0.7t}$. Therefore the function $\tilde{S}$ *is* a solution to the differential equation. Let's build off this to try other candidate functions:

Example 7.1. Verify if the following functions are solutions to Equation (7.1):

- $\tilde{R}(t) = 10e^{0.7t}$
- $\tilde{P}(t) = e^{0.7t}$

- $\tilde{Q}(t) = 5e^{0.7t}$
- $\tilde{F}(t) = 3$
- $\tilde{G}(t) = 0$

Solution. First apply direct differentiation to each of these functions (this represents the left hand side of each differential equation):

- $\tilde{R}(t) = 10e^{0.7t} \to \tilde{R}'(t) = 7e^{0.7t}$
- $\tilde{P}(t) = e^{0.7t} \to \tilde{P}'(t) = 0.7e^{0.7t}$
- $\tilde{Q}(t) = 5e^{0.7t} \to \tilde{Q}'(t) = 3.5e^{0.7t}$
- $\tilde{F}(t) = 3 \to \tilde{F}'(t) = 0$
- $\tilde{G}(t) = 0 \to \tilde{G}'(t) = 0$

Next compare each of these solutions to the right hand side of Equation (7.1):

- $0.7\tilde{R}(t) = 0.7 \cdot 10e^{0.7t} \to 7e^{0.7t}$
- $0.7\tilde{P}(t) = 0.7e^{0.7t}$
- $0.7\tilde{Q}(t) = 0.7 \cdot 5e^{0.7t} \to= 3.5e^{0.7t}$
- $0.7\tilde{F}(t) = 0.7 \cdot 3 \to 2.1$
- $0.7\tilde{G}(t) = 0.7 \cdot 0 \to 0$

Notice how in the candidate functions (with the exception of $\tilde{F}(t)$) the right hand side of each equation equals the left hand side. When that is the case, our candidate functions are indeed solutions to Equation (7.1)!

Verifying a solution to a differential equation relies on your knowledge of differentiation versus other more involved methods, which may be an under-appreciated approach. In some instances you may not be given a candidate function as in Example 7.1. Deciding "what function to try" is the hardest step, but a safe bet would be an exponential equation (especially if the right hand side involves the dependent variable). As we will see in Chapter 18 the guess and check approach will help us to determine general solutions to systems of linear differential equations.

7.1.1 Superposition of solutions

Related to the verification method is a concept called superposition of solutions. Here is how this works: if you have two known solutions to a differential equation, then the sum (or difference) is a solution as well. Let's look at an example:

Example 7.2. Show that $\tilde{R}(t) + \tilde{Q}(t) = 5e^{0.7t} + e^{0.7t}$ from Example 7.1 is a solution to Equation (7.1).

Solution. By direct differentiation, $\tilde{R}'(t) + \tilde{Q}'(t) = 3.5e^{0.7t} + 0.7e^{0.7t}$. Further-more, $0.7 \cdot (\tilde{R}(t) + \tilde{Q}(t)) = 0.7 \cdot (5e^{0.7t} + e^{0.7t}) = 3.5e^{0.7t} + 0.7e^{0.7t}$, which equals $\tilde{R}'(t) + \tilde{Q}'(t)$.

Example 7.2 illustrates the principle that different solutions to a differential equation can be added together and produce a new solution. More generally, adding two solutions together is an example of a *linear combination* of solutions, and we can state this more formally:

If $x(t)$ and $y(t)$ are solutions to the differential equation $z' = f(t, z)$, then $c(t) = a \cdot x(t) + b \cdot y(t)$ are also solutions, where a and b are constants.

7.2 Separable differential equations

The next techninque is called *separation of variables*. This method has a defined workflow (Separate $\rightarrow$ Integrate $\rightarrow$ Solve), which we illustrate by considering the following differential equation:

$$\frac{dy}{dt} = yt^2 \tag{7.2}$$

Separate: This step uses algebra to collect variables involving x on one side of the equation, and the variables involving y on the other (Equation (7.3)):

$$\frac{1}{y}dy = t^2 \, dt \tag{7.3}$$

Integrate: This step computes the antiderivative of both sides of Equation (7.3):

$$\int \frac{1}{y}dy = \ln(y) + C_1.$$
$$\int t^2 \, dt = \frac{1}{3}t^3 + C_2 \tag{7.4}$$

You may remember from calculus that whenever you compute antiderivatives to always include a $+C$ (hence the C_1 and C_2 in Equation (7.4)). For separable differential equations it is okay just to keep only one of the $+C$ terms in Equation (7.4), which usually is best on the side of the independent variable (in this case t). Since both sides of the separated equation are equal, we can rewrite Equation (7.4) on a single line (Equation (7.5)):

$$\ln(y) = \frac{1}{3}t^3 + C \tag{7.5}$$

Solve: This last step solves Equation (7.5) for the dependent variable y:

$$\ln(y) = \frac{1}{3}t^3 + C \rightarrow e^{\ln(y)} = e^{\frac{1}{3}t^3 + C} = e^C \cdot e^{\frac{1}{3}t^3} \rightarrow y = Ce^{\frac{1}{3}t^3} \tag{7.6}$$

We are in business! Notice how in Equation (7.6) at each step we just kept the constant to be C, since exponentiating a constant will still be constant.

To summarize, the workflow for the separating of variables technique is the following:

1. **Separate** the variables on one side of the equation.
2. **Integrate** both sides individually.
3. **Solve** for the dependent variable.

7.3   Integrating factors

Chapters 1 and 4 examined a model for the spread of Ebola where that was proportional to the number infected:

$$\frac{dI}{dt} = .023(13600 - I) = 312.8 - .023I \tag{7.7}$$

While Equation (7.7) can be solved via separation of variables, let's try a different approach to illustrate another useful technique. First let's write the terms in Equation (7.7) that involve I on one side of the equation:

$$\frac{dI}{dt} + .03I = 30. \tag{7.8}$$

What we are going to do is multiply both sides of this Equation (7.8) by $e^{.023t}$ (I'll explain more about that later):

$$\frac{dI}{dt} \cdot e^{.023t} + .023I \cdot e^{.023t} = 312.8 \cdot e^{.023t} \tag{7.9}$$

Hmmm - this seems like we are making the differential equation harder to solve, doesn't it? However the left hand side of Equation (7.9) is actually the derivative of the expression $I \cdot e^{.023t}$ (courtesy of the product rule from calculus). Let's take a look:

$$\frac{d}{dt}\left(I \cdot e^{.023t}\right) = \frac{dI}{dt} \cdot e^{.023t} + I \cdot .023e^{.023t} \tag{7.10}$$

Equation (7.10) allows us to express the left hand side of Equation (7.9) as a derivative and then integrate both sides:

$$\frac{d}{dt}\left(I \cdot e^{.023t}\right) = 312.8 \cdot e^{.023t} \rightarrow$$

$$\int \frac{d}{dt}\left(I \cdot e^{.023t}\right) \, dt = \int 312.8 \cdot e^{.023t} \, dt \rightarrow \qquad (7.11)$$

$$I \cdot e^{.023t} = 13600 \cdot e^{.023t} + C$$

All that is left to do is to solve Equation (7.11) in terms of $I(t)$ by dividing by $e^{.023t}$, labeled as $I_1(t)$ (Equation (7.12)):

$$I_1(t) = 13600 + Ce^{-.023t} \qquad (7.12)$$

Cool! The function $f(t) = e^{.023t}$ is called an *integrating factor*. Let's explore this technique with a second example:

Example 7.3. Apply the integrating factor technique to determine a general solution to the differential equation:

$$\frac{dI}{dt} = .023t(13600 - I) = 312.8t - .023t \cdot I \qquad (7.13)$$

(Equation (7.13) is a modification of Equation (7.7), where the rate of infection is time dependent.)

Solution. Re-writing Equation (7.13) we have:

$$\frac{dI}{dt} + .023t \cdot I = 312.8t \qquad (7.14)$$

To write the left hand side of Equation (7.14) as the derivative of a product of functions, multiply the *entire* differential equation by $e^{\int .023t \, dt} = e^{.0115t^2}$. This term is called the *integrating factor* (Equation (7.15)):

$$\frac{dI}{dt} \cdot e^{.0115t^2} + .023t \cdot I \cdot e^{0.0115t^2} = 312.8t \cdot e^{0.0115t^2} \qquad (7.15)$$

Rewrite the left hand side of Equation (7.15) with the product rule:

$$\frac{dI}{dt} \cdot e^{0.0115t^2} + .023t \cdot I \cdot e^{0.0115t^2} = \frac{d}{dt}\left(I \cdot e^{0.0115t^2}\right) \qquad (7.16)$$

Next integrate Equation (7.16):

$$\frac{d}{dt}\left(I \cdot e^{.0115t^2}\right) = 312.8t \cdot e^{.0115t^2} \rightarrow$$

$$\int \frac{d}{dt}\left(I \cdot e^{.0115t^2}\right) \, dt = \int 312.8t \cdot e^{.0115t^2} \, dt \rightarrow \qquad (7.17)$$

$$I \cdot e^{0.0115t^2} = 27200 \cdot e^{.0115t^2} + C$$

The final step is to write the Equation (7.17) in terms of $I(t)$; we will label this solution as $I_2(t)$:

$$I_2(t) = 27200 + Ce^{-.0115t^2} \qquad (7.18)$$

Figure 7.1 compares the solutions $I_1(t)$ and $I_2(t)$ when the initial condition (in both cases) is 10 (so $I_1(0) = I_2(0) = 10$). Notice how the extra time-dependent factor in Equation (7.13) makes the cases grow quickly before leveling off.

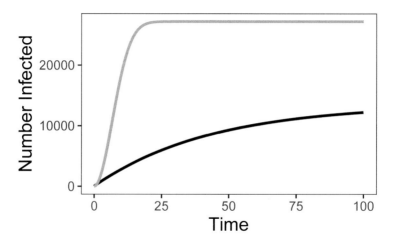

FIGURE 7.1 Comparison of two integrating factor solutions, Equation (7.12) in red and Equation (7.18) in blue.

The integrating factor approach can be applied for differential equations that can be written in the form $\frac{dy}{dt} + f(t) \cdot y = g(t)$, using the following workflow:

1. **Determine** the *integrating factor* $e^{\int f(t) \, dt}$. Hopefully the integral $\int f(t) \, dt$ is easy to compute!

2. **Multiply** the integrating factor across your equation to rewrite the differential equation as $\frac{d}{dt}\left(y \cdot e^{\int f(t) \, dt}\right) = g(t) \cdot e^{\int f(t) \, dt}$.

3. **Evaluate** the integral $H(t) = \int g(t) \cdot e^{\int f(t)\,dt}\,dt$. This looks intimidating - but hopefully is manageable to compute! Don't forget the $+C$!

4. **Solve** for $y(t)$: $y(t) = H(t) \cdot e^{-\int f(t)\,dt} + Ce^{-\int f(t)\,dt}$.

7.4 Applying the verification method to coupled equations

Finally, the method of verifying a solution helps to introduce a useful solution technique for systems of differential equations. We are going to study a simplified version of the lynx-hare model from Chapter 6. Equation (7.19) assumes both lynx and hares decline at a rate proportional to their respective population sizes, with the decline in the lynx population representing predation:

$$\frac{dH}{dt} = -bH$$
$$\frac{dL}{dt} = bH - dL \tag{7.19}$$

Based on these simplified assumptions a good approach is to assume a solution that is exponential for both H and L (Equation (7.20)):

$$\tilde{H}(t) = C_1 e^{\lambda t}$$
$$\tilde{L}(t) = C_2 e^{\lambda t} \tag{7.20}$$

Notice that Equation (7.20) has C_1, C_2, and λ as parameters.[1] Let's apply the verification method to determine expressions for C_1, C_2, and λ that are consistent with Equation (7.19). By differentiation of Equation (7.20), we have the following (Equation (7.21)):

$$\frac{d\tilde{H}}{dt} = \lambda C_1 e^{\lambda t}$$
$$\frac{d\tilde{L}}{dt} = \lambda C_2 e^{\lambda t} \tag{7.21}$$

Comparing Equation (7.21) to Equation (7.19) we can show that

$$\lambda C_1 e^{\lambda t} = -bC_1 e^{\lambda t}$$
$$\lambda C_2 e^{\lambda t} = bC_1 e^{\lambda t} - dC_2 e^{\lambda t} \tag{7.22}$$

---

[1] If you have taken a course in linear algebra, you may recognize that we are assuming the solution is a vector of the form $\vec{v} = \vec{C}e^{\lambda t}$.

Let's rearrange Equation (7.21) some more:

$$(\lambda + b)C_1 e^{\lambda t} = 0$$
$$(\lambda + d)C_2 e^{\lambda t} = bC_1 e^{\lambda t} \qquad (7.23)$$

Notice that for the second expression in Equation (7.23) we can express $C_1 e^{\lambda t}$ as $\dfrac{(\lambda + d)}{b} C_2 e^{\lambda t}$. Next, we can substitute this expression for $C_1 e^{\lambda t}$ into the first expression of Equation (7.23):

$$(\lambda + b)\frac{(\lambda + d)}{b} C_2 e^{\lambda t} = 0 \qquad (7.24)$$

If we assume that $b \neq 0$, then we have the following simplified expression (Equation (7.25)):

$$(\lambda + b)(\lambda + d)C_2 e^{\lambda t} = 0 \qquad (7.25)$$

Because the exponential function in Equation (7.25) never equals zero, the only possibility is that $(\lambda + b)(\lambda + d) = 0$, or that $\lambda = -b$ or $\lambda = -d$.[2]

Next we need to determine values of C_1 and C_2. We can do this by going back to the the second expression in Equation (7.22), which we rearrange to Equation (7.26):

$$(\lambda + d)C_2 e^{\lambda t} - bC_1 e^{\lambda t} = 0 \qquad (7.26)$$

Let's analyze Equation (7.26) for each of the values of λ:

- Case $\lambda = -d$
  For this case we have Equation (7.27):

$$(-d + d)C_2 e^{-dt} - bC_1 e^{-dt} = 0 \rightarrow -bC_1 e^{-dt} = 0 \qquad (7.27)$$

The only way for Equation (7.27) to be consistent and remain zero is if $C_1 = 0$. We don't have any restrictions on C_2, so the general solution is given with Equation (7.28):

$$\tilde{H}(t) = 0$$
$$\tilde{L}(t) = C_2 e^{-dt} \qquad (7.28)$$

- Case $\lambda = -d$ For this situation, Equation (7.26) becomes Equation (7.29):

---

[2]Remember: if expressions multiply to zero, then the only possibility is that at least one of them is zero. The process outlined here finds the *eigenvalues* and *eigenvectors* of a system of equations. We will study these concepts in Chapter 18.

$$((-d + b)C_2 - dC_1) e^{-dt} = 0 \tag{7.29}$$

The only way for Equation (7.29) to be consistent is if $((-d + b)C_2 - dC_1) = 0$, or if $C_2 = \left(\dfrac{d}{-d + b} \right) C_1$. In this case, Equation (7.30) then represents the general solution:

$$\tilde{H}(t) = C_1 e^{-dt}$$
$$\tilde{L}(t) = \left(\frac{d}{-d + b} \right) C_1 e^{-dt} \tag{7.30}$$

The parameter C_2 in Equation (7.30) can be determined by the initial condition. Notice that we need to have $d \neq b$ or our solution will be undefined.

Finally we can write down a general solution to Equation (7.19) by combining our Equations (7.28) and (7.30) by superposition (Equation (7.31)):

$$H(t) = C_1 e^{-dt}$$
$$L(t) = \left(\frac{d}{-d + b} \right) C_1 e^{-dt} + C_2 e^{-bt} \tag{7.31}$$

The method outlined here only works on *linear* differential equations (i.e. it wouldn't work if there was a term such as kHL in Equation (7.19)). In Chapter 18 explores this method more systematically to determine general solutions to linear systems of equations.

As you can see, there are a variety of techniques that can be applied in the solution of differential equations. Many more solution techniques exist - but by and large the techniques presented here probably will be your "go-tos" when working to find an exact solution to a differential equation.

7.5   Exercises

Exercise 7.1. Determine the value of C when $I(0) = 10$ for the two equations:

$$I_1(t) = 1000 + Ce^{-.03t}$$
$$I_2(t) = 1000 + Ce^{-0.015t^2} \tag{7.32}$$

Exercise 7.2. Solve Equation (7.7) using the separation of variables technique.

Exercise 7.3. Verify that $I_2(t) = N + Ce^{-0.5kt^2}$ is the solution to the differential equation $\dfrac{dI}{dt} = kt(N - I)$. Set $N = 3$ and $C = 1$. Plot $I_2(t)$ with various values of k ranging from .001 to .1. What effect does k have on the solution?

Exercise 7.4. Consider the following model of an infection:

$$\frac{dS}{dt} = -kSI$$

$$\frac{dI}{dt} = kSI - \beta I \qquad (7.33)$$

Use this equation to solve the following questions:

a. Show that $\dfrac{I'}{S'} = -1 + \dfrac{\beta}{k}\dfrac{1}{S}$, where $S' = \dfrac{dS}{dt}$ and $I' = \dfrac{dI}{dt}$. We will call $\dfrac{I'}{S'} = \dfrac{dI}{dS}$.

b. Using separation of variables, show that the general solution to $\dfrac{I'}{S'} = -1 + \dfrac{\beta}{k}\dfrac{1}{S}$ is $I(S) = -S + \dfrac{\beta}{k}\ln(S) + C$.

c. At the beginning of the epidemic, $S_0 + I_0 = N$, where N is the total population size. Use this fact to determine C in the equation $I_0 = -S_0 + \dfrac{\beta}{k}\ln(S_0) + C$.

d. Using your previous answer, show that $I(S) = N - S + \dfrac{\beta}{k}\ln\left(\dfrac{S}{S_0}\right)$.

e. Plot a solution curve for $I(S)$ with $\beta = 1$, $k = 0.1$, $N = 100$, and $S_0 = 5$.

Exercise 7.5. (Inspired by Scholz and Scholz (2014)) A chemical reaction $2A \rightarrow C + D$ can be modeled with the following differential equation:

$$\frac{dA}{dt} = -2kA^2 \qquad (7.34)$$

Apply the method of separation of variables to determine a general solution for this differential equation.

Exercise 7.6. (Inspired by Hugo van den Berg (2011)) Organisms that live in a saline environment biochemically maintain the amount of salt in their blood stream. An equation that represents the level of S in the blood is the following:

$$\frac{dS}{dt} = I + p \cdot (W - S) \qquad (7.35)$$

where the parameter I represents the active uptake of salt, p is the permeability

of the skin, and W is the salinity in the water. For this problem, set $I = 0.1$
(10% / hour), $p = 0.05 \text{ hr}^{-1}$, $W = 0.4$ (40% salt concentration), and $S(0) = 0.6$
(60% salt concentration).

a. Generate a slope field of the differential equation $\dfrac{dS}{dt} = 0.1 + 0.05 \cdot (.6 - S)$.

b. Apply integrating factors to solve the differential equation $\dfrac{dS}{dt} = 0.1 +$
   $0.05 \cdot (.6 - S)$.

c. Does your solution conform to the slope field diagram?

Exercise 7.7. Which of the following differential equations can be solved via
separation of variables?

a. $\dfrac{dy}{dx} = x^2 + xy$

b. $\dfrac{dy}{dx} = e^{x+y}$

c. $\dfrac{dy}{dx} = y \cdot \cos(2 + x)$

d. $\dfrac{dy}{dx} = \ln x + \ln y$

e. $\dfrac{dy}{dx} = x \cdot (y^2 + 2)$

Once you have identified which ones can be solved via separation of variables,
apply that technique to solve each differential equation.

Exercise 7.8. Consider the following differential equation $\dfrac{dP}{dt} = -\delta P$, $P(0) =$
P_0, where δ is a constant parameter.

a. Solve this equation using the method of separation of variables.
b. Solve this equation using an integrating factor.
c. Your two solutions from the two methods should be the same - are they?

Exercise 7.9. A differential equation that relates a consumer's nutrient
content (denoted as y) to the nutrient content of food (denoted as x) is given
by:

$$\frac{dy}{dx} = \frac{1}{\theta}\frac{y}{x}, \tag{7.36}$$

where $\theta \geq 1$ is a constant. Apply separation of variables to determine the
general solution to this differential equation.

Exercise 7.10. Apply separation of variables to determine general solutions
to the following systems of differential equations:

$$\frac{dx}{dt} = x$$
$$\frac{dy}{dt} = y$$
(7.37)

(Equation (7.37) is an example of an *uncoupled* system of equations.)

Exercise 7.11. (Inspired by Thornley and Johnson (1990)) A plant grows proportional to its current length L. Assume this proportionality constant is μ, whose rate also decreases proportional to its current value. The system of equations that models this plant growth is the following:

$$\frac{dL}{dt} = \mu L$$
$$\frac{d\mu}{dt} = -k\mu$$
(7.38)

(k is a constant parameter)

Apply separation of variables to determine the general solutions to this system of equations.

Exercise 7.12. Apply the verification method developed in Chapter 7.4 to determine the general solution to the following system of differential equations:

$$\frac{dx}{dt} = x - y$$
$$\frac{dy}{dt} = 2y$$
(7.39)

Exercise 7.13. Apply the integrating factors technique to determine the solution to the differential equation $\frac{dI}{dt} = (N - I) = kN - kI$, where k and N are parameters.

Exercise 7.14. For each of the following differential equations:

- Determine equilibrium solutions for the differential equation.
- Apply separation of variables to determine general solutions to the following differential equations.
- Choose reasonable values of any parameters and plot the solution curve for an initial condition that you select.

a. $\dfrac{dy}{dx} = -\dfrac{x}{y}$

b. $\dfrac{dy}{dx} = 8 - y$

c. $\dfrac{dW}{dt} = k(N - W)$ (k and N are constant parameters)

d. $\dfrac{dR}{dt} = -aR\ln\dfrac{R}{K}$ (a and K are constant parameters)

Exercise 7.15. Consider the following differential equation, where M represents a population of mayflies and t is time (given in months), and δ is a mortality rate (units % mayflies / month):

$$\frac{dM}{dt} = -\delta \cdot M \qquad (7.40)$$

Determine the general solution to this differential equation and plot a few different solution curves with different values of $\delta > 0$. Assume that $M(0) = 10,000$. Describe the effect of changing δ on your solution.

Exercise 7.16. An alternative model of mayfly mortality is the following:

$$\frac{dM}{dt} = -\delta(t) \cdot M, \qquad (7.41)$$

where $\delta(t)$ is a time dependent mortality function. Determine a solution and plot a solution curve (assuming $M(0) = 10,000$ and over the interval from $0 \le t \le 5$, assuming time is scaled appropriately) for this differential equation when $\delta(t)$ has the following forms:

a. $\delta(t) = 1$
b. $\delta(t) = 2t$
c. $\delta(t) = 1 - e^{-t}$
d. $\delta(t) = 1 + e^{-t}$

Provide a reasonable biological explanation justifying the use of these alternative mayfly models.

8

Linear Regression and Curve Fitting

8.1 What is parameter estimation?

Chapters 1 - 7 introduced the idea of modeling with rates of change. There is much more to be stated regarding qualitative analyses of a differential equation (Chapters 5 and 6), which we will return to starting in Chapter 15. But for the moment, let's pause and recognize that a key motivation for modeling with rates of change is to quantify observed phenomena.

Oftentimes, we wish to compare model outputs to measured data. While that may seem straightforward, sometimes models have parameters (such as k and β for Equation (6.1) in Chapter 6). Parameter estimation is the process of determining model parameters from data. Stated differently:

> **Parameter estimation** is the process that determines the set of parameters $\vec{\alpha}$ that minimize the difference between data $\vec{y}$ and the output of the function $f(\vec{x}, \vec{\alpha})$ and measured error $\vec{\sigma}$.

Over the next several chapters we will examine aspects of *parameter estimation*. Sometimes parameter estimation is synonymous with "fitting a model to data" and can also be called *data assimilation* or *model data fusion*. We can address the parameter estimation problem from several different mathematical areas: *calculus* (optimization), *statistics* (likelihood functions), and *linear algebra* (systems of linear equations). We will explore how we define "best" over several chapters, but let's first explore techniques of how this is done in R using simple linear regression.[1] Let's get started!

---

[1]We will use R a lot in this chapter to make plots - so please visit Chapter 2 if you need some reminders on plotting in R.

TABLE 8.1 First few years of average global temperature anomalies. The anomaly represents the global surface temperature relative to 1951-1980 average temperatures.

year_since_1880	0.00	1.00	2.00	3.00	4.00	5.00	6.00
temperature_anomaly	-0.17	-0.08	-0.11	-0.18	-0.29	-0.33	-0.31

8.2 Parameter estimation for global temperature data

Let's take a look at a specific example. Table 8.1 shows anomalies in average global temperature since 1880, relative to 1951-1980 global temperatures.[2] This dataset can be found in the demodelr package with the name global_temperature. To name our variables let $Y = $ Year since 1880 and $T = $ Temperature anomaly.

We will be working with these data to fit a function $f(Y, \vec{\alpha}) = T$. In order to fit a function in R we need three essential elements, distilled into a workflow of: Identify → Construct → Compute

- **Identify** data for the formula to estimate parameters. For this example we will use the tibble (or data frame) global_temperature.
- **Construct** the regression formula we will use for the fit. We want to do a linear regression so that $T = a + bY$. How we represent the regression formula in R is with temperature_anomaly ~ 1 + year_since_1880. Notice that this regression formula must include *named columns from your data*. Said differently, this regression formula "defines a linear regression where the factors are a constant term and one is proportional to the predictor variable." It is helpful to assign this regression formula as a variable: regression_formula <- temperature_anomaly ~ 1 + year_since_1880. In Chapter 8.3 we will discuss other types of regression formulas.
- **Compute** the regression with the **command** lm (which stands for *l*inear *m*odel).

That's it! So if we need to do a linear regression of global temperature against year since 1880, it can be done with the following code:

```r
regression_formula <- temperature_anomaly ~ 1 + year_since_1880

linear_fit <- lm(regression_formula, data = global_temperature)

summary(linear_fit)

##
```

---

[2]Data provided by NOAA: https://climate.nasa.gov/vital-signs/global-temperature/.

```
## Call:
## lm(formula = regression_formula, data = global_temperature)
##
## Residuals:
##      Min       1Q    Median       3Q      Max
## -0.35606 -0.13185 -0.03044  0.12449  0.45518
##
## Coefficients:
##                   Estimate Std. Error t value Pr(>|t|)
## (Intercept)     -0.4880971  0.0296270  -16.48   <2e-16 ***
## year_since_1880  0.0076685  0.0003633   21.11   <2e-16 ***
## ---
## Signif. codes:  0 '***' 0.001 '**' 0.01 '*' 0.05 '.' 0.1 ' ' 1
##
## Residual standard error: 0.1775 on 140 degrees of freedom
## Multiple R-squared:  0.7609, Adjusted R-squared:  0.7592
## F-statistic: 445.6 on 1 and 140 DF,  p-value: < 2.2e-16
```

What is printed on the console (and shown above) is the summary of the fit
results. This summary contains several interesting things that you would study
in advanced courses in statistics, but here is what we will focus on:

- The estimated **coefficients** (starting with `Coefficients:` above) of the
  linear regression. The column `Estimate` lists the constants in front of our
  regression formula $y = a + bx$. What follows is the statistical error for that
  estimate. The other additional columns concern statistical tests that show
  significance of the estimated parameters.
- One helpful thing to look at is the **Residual standard error** (starting
  with `Residual standard error` above), which represents the overall, total
  effect of the differences between the model predicted values of $\vec{y}$ and the
  measured values of $\vec{y}$. The goal of linear regression is to minimize this
  model-data difference.

The summary of the statistical fit is a verbose readout, which may prohibit
quickly identifying the regression coefficients or plotting the fitted results.
Thankfully the `R` package called `broom` can help us! The `broom` package produces
model output in what is called "tidy" data format. You can read more about
`broom` from its documentation.[3]

Since we are only going to use one or two functions from this package, I am
going to refer to the functions I need with the syntax `PACKAGE_NAME::FUNCTION`.

First we will make a data frame with the predicted coefficients from our linear
model, as shown with the following code that you can run on your own:

---

[3] https://broom.tidymodels.org/index.html

```
global_temperature_model <-
  broom::augment(linear_fit, data = global_temperature)

glimpse(global_temperature_model)
```

Notice how the augment command takes the results from linear_fit with the data global_temperature to produce model estimated results (under the variable named .fitted.[4] There is a lot to unpack with this new data frame, but the important ones are the columns year_since_1880 (the independent variable) and .fitted, which represents the fitted coefficients.

Finally, Figure 8.1 compares the data to the fitted regression line (also known as the "best fit line").

```
ggplot(data = global_temperature) +
  geom_point(aes(x = year_since_1880, y = temperature_anomaly),
    color = "red",
    size = 2
  ) +
  geom_line(
    data = global_temperature_model,
    aes(x = year_since_1880, y = .fitted)
  ) +
  labs(
    x = "Year Since 1880",
    y =  "Temperature anomaly"
  )
```

---

[4]I like appending _model to the original name of the data frame to signify we are working with modeled components.

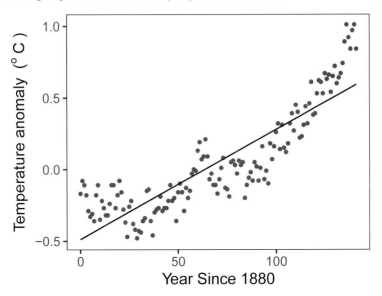

FIGURE 8.1 Global temperature anomaly data along with the fitted linear regression line.

A word of caution: this example is one of an exploratory data analysis illustrating parameter estimation with R. Global temperature is a measurement of many different types of complex phenomena integrated from the local, regional, continental, and global levels (and interacting in both directions). Global temperature anomalies cannot be distilled down to a simple linear relationship between time and temperature. What Figure 8.1 *does* suggest though is that over time the average global temperature has increased. I encourage you to study the pages at NOAA[5] to learn more about the scientific consensus in modeling climate (and the associated complexities - it is a fascinating scientific problem that will need YOUR help to solve it!)

8.3 Moving beyond linear models for parameter estimation

We can also estimate parameters or fit additional polynomial models such as the equation $y = a + bx + cx^2 + dx^3$... (here the estimated parameters a, b, c, d, ...). There is a key distinction here: the regression formula is *nonlinear* in the predictor variable x, but *linear* with respect to the parameters. Incorporating

---

[5] https://climate.nasa.gov/

these regression formulas in R modifies the structure of the regression formula. A few templates are show in Table 8.2:

TABLE 8.2 Comparison of model equations to regression formulas used for R. The variable y is the response variable and x the predictor variable. Notice the structure I(..) is needed for R to signify a factor of the form x^n.

Equation	Regression Formula
$y = a + bx$	y ~ 1 + x
$y = a$	y ~ 1
$y = bx$	y ~ -1+x
$y = a + bx + cx^2$	y ~ 1 + x + I(x^2)
$y = a + bx + cx^2 + dx^3$	y~ 1 + x + I(x^2) + I(x^3)

8.3.1   Can you linearize your model?

We can estimate parameters for nonlinear models in cases where the function can be transformed mathematically to a linear equation. Here is one example: while the equation $y = ae^{bx}$ is nonlinear with respect to the parameters, it can be made linear by a *logarithmic transformation* of the data:/index%7Blogarit hmic transformation}

$$\ln(y) = \ln(ae^{bx}) = \ln(a) + \ln(e^{bx}) = \ln(a) + bx \tag{8.1}$$

The advantage to this approach is that the growth rate parameter b is easily identifiable from the data, and the value of a is found by exponentiation of the fitted intercept value. The disadvantage is that you need to do a log transform of the y variable first before doing any fits.

Example 8.1. A common nonlinear equation in enzyme kinetics is the *Michaelis-Menten* law, which states that the rate of the uptake of a substrate V is given by the equation:

$$V = \frac{V_{max}s}{s + K_m}, \tag{8.2}$$

where s is the amount of substrate, K_m is half-saturation constant, and V_{max} the maximum reaction rate. (Typically V is used to signify the "velocity" of the reaction.)

Consider you have the following data (from J. Keener and Sneyd (2009)):

s (mM)	0.1	0.2	0.5	1.0	2.0	3.5	5.0
V (mM / s)	0.04	0.08	0.17	0.24	0.32	0.39	0.42

Apply parameter estimation techniques to estimate K_m and V_{max} and plot the resulting fitting curve with the data.

Solution. The first thing that we will need to do is to define a data frame (`tibble`) for these data:

```
enzyme_data <- tibble(
  s = c(0.1, 0.2, 0.5, 1.0, 2.0, 3.5, 5.0),
  V = c(0.04, 0.08, 0.17, 0.24, 0.32, 0.39, 0.42)
)
```

Figure 8.2 shows a plot of s and V:

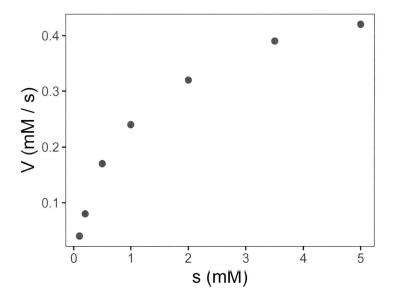

FIGURE 8.2 Scatterplot of enzyme substrate data from Example 8.1.

Figure 8.2 definitely suggests a nonlinear relationship between s and V. To dig a little deeper, try running the following code that plots the reciprocal of s and the reciprocal of V (do this on your own):

```
ggplot(data = enzyme_data) +
  geom_point(aes(x = 1 / s, y = 1 / V),
```

```
    color = "red",
    size = 2
  ) +
  labs(
    x = "1/s (1/mM)",
    y = "1/V (s / mM)"
  )
```

Notice how easy it was to plot the reciprocals of s and V inside the ggplot command. Here's how to see this with the provided equation (and a little bit of algebra):

$$V = \frac{V_{max}s}{s + K_m} \rightarrow \frac{1}{V} = \frac{s + K_m}{V_{max}s} = \frac{1}{V_{max}} + \frac{1}{s}\frac{K_m}{V_{max}} \tag{8.3}$$

In order to do a linear fit to the transformed data we will use the regression formulas defined above and the handy structure I(VARIABLE) and plot the transformed data with the fitted equation (do this on your own as well):[6]

```
# Define the regression formula
enzyme_formula <- I(1 / V) ~ 1 + I(1 / s)

# Apply the linear fit
enzyme_fit <- lm(enzyme_formula,data = enzyme_data)

# Show best fit parameters
summary(enzyme_fit)

# Added fitted data to the model
enzyme_data_model <- broom::augment(enzyme_fit, data = enzyme_data)

# Compare fitted model to the data
ggplot(data = enzyme_data) +
  geom_point(aes(x = 1 / s, y = 1 / V),
    color = "red",
    size = 2
  ) +
  geom_line(
    data = enzyme_data_model,
    aes(x = 1 / s, y = .fitted)
  ) +
  labs(
    x = "1/s (1/mM)",
```

---

[6]The process outlined here forms a *Lineweaver-Burk* plot.

```
  y = "1/V (s / mM)"
)
```

In Exercise 8.6 you will use the coefficients from your linear fit to determine V_{max} and K_m. When plotting the fitted model values with the original data (Figure 8.3), we need to take the reciprocal of the column .fitted when we apply augment because the response variable in the linear model is $1/V$ (confusing, I know!). For convenience, the code that does the all the fitting is shown below:[7]

```
# Define the regression formula
enzyme_formula <- I(1 / V) ~ 1 + I(1 / s)

# Apply the linear fit
enzyme_fit <- lm(enzyme_formula,data = enzyme_data)

# Added fitted data to the model
enzyme_data_model <- broom::augment(enzyme_fit, data = enzyme_data)


ggplot(data = enzyme_data) +
  geom_point(aes(x = s, y = V),
    color = "red",
    size = 2
  ) +
  geom_line(
    data = enzyme_data_model,
    aes(x = s, y = 1 / .fitted)
  ) +
  labs(
    x = "s (mM)",
    y = "V (mM / s)"
  )
```

---

[7]You may notice that in Figure 8.3 the fitted curve seems to look like a piecewise linear function. This is mainly due to the distribution of data - if you have several gaps between measurements, the fitted curve looks smoother.

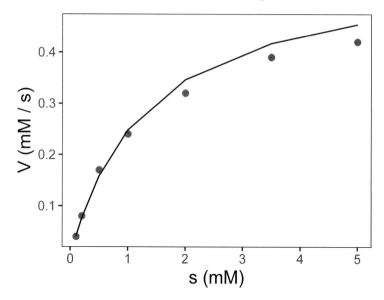

FIGURE 8.3 Scatterplot of enzyme substrate data from Example 8.1 along with the fitted curve.

8.4 Parameter estimation with nonlinear models

In many cases you will not be able to write your model in a linear form by applying functional transformations. Here's the good news: you can still do a non-linear curve fit using the function nls, which is similar to the command lm with some additional information. Let's return to an example from Chapter 2 where we examined the weight of a dog named Wilson (Figure 2.2). One model for the dog's weight W from the days since birth D is a saturating exponential equation (Equation (8.4)):

$$W = f(D, a, b, c) = a - be^{-ct},\qquad(8.4)$$

where we have the parameters a, b, and c. We can apply the command nls (nonlinear least squares) to estimate these parameters. Because nls is an iterative numerical method it needs a starting value for these parameters (here we set $a = 75$, $b = 30$, and $c = 0.01$). Determining a starting point can be tricky - it does take some trial and error.

```
# Define the regression formula
wilson_formula <- weight ~ a - b * exp(-c * days)
```

```r
# Apply the nonlinear fit
nonlinear_fit <- nls(formula = wilson_formula,
  data = wilson,
  start = list(a = 75, b = 30, c = 0.01)
)

# Summarize the fit parameters
summary(nonlinear_fit)

##
## Formula: weight ~ a - b * exp(-c * days)
##
## Parameters:
##     Estimate Std. Error t value Pr(>|t|)
## a 7.523e+01  1.573e+00   47.84  < 2e-16 ***
## b 9.077e+01  3.400e+00   26.70 1.07e-14 ***
## c 6.031e-03  4.324e-04   13.95 2.26e-10 ***
## ---
## Signif. codes:  0 '***' 0.001 '**' 0.01 '*' 0.05 '.' 0.1 ' ' 1
##
## Residual standard error: 2.961 on 16 degrees of freedom
##
## Number of iterations to convergence: 11
## Achieved convergence tolerance: 6.899e-06
```

Similar as before, we can augment the data to display the fitted curve.

However once you have your fitted model, you can still plot the fitted values with the coefficients (try this out on your own).

```r
# Augment the model
wilson_model <- broom::augment(nonlinear_fit, data = wilson)

# Plot the data with the model
ggplot(data = wilson) +
  geom_point(aes(x = days, y = weight),
    color = "red",
    size = 2
  ) +
  geom_line(
    data = wilson_model,
    aes(x = days, y = .fitted)
  ) +
  labs(
    x = "Days since birth",
```

```
    y = "Weight (pounds)"
  )
```

8.5   Towards model-data fusion

The R language (and associated packages) has many excellent tools for parameter estimation and comparing fitted models to data. These tools are handy for first steps in parameter estimation.

More broadly the technique of estimating models from data can also be called *data assimilation* or *model-data fusion*. Whatever terminology you happen to use, you are combining the best of both worlds: combining observed measurements with what you expect *should* happen, given the understanding of the system at hand.

We are going to dig into data assimilation even more - and one key tool is understanding likelihood functions, which we will study in the next chapter.

8.6   Exercises

Exercise 8.1. Determine if the following equations are linear with respect to the parameters. Assume that y is the response variable and x the predictor variable.

a. $y = a + bx + cx^2 + dx^3$
b. $y = a\sin(x) + b\cos(x)$
c. $y = a\sin(bx) + c\cos(dx)$
d. $y = a + bx + a \cdot bx^2$
e. $y = ae^{-x} + be^x$
f. $y = ae^{-bx} + ce^{-dx}$

Exercise 8.2. Each of the following equations can be written as linear with respect to the parameters, through applying some elementary transformations to the data. Write each equation as a linear function with respect to the parameters. Assume that y is the response variable and x the predictor variable.

a. $y = ae^{-bx}$
b. $y = (a + bx)^2$
c. $y = \dfrac{1}{a + bx}$
d. $y = cx^n$

Exercise 8.3. Use the dataset `global_temperature` and the function `lm` to answer the following questions:

a. Complete the following table, which represents various regression fits to global temperature anomaly T (in degrees Celsius) and years since 1880 (denoted by Y). In the table **Coefficients** represent the values of the parameters a, b, c, etc. from your fitted equation; **P** = number of parameters; **RSE** = Residual standard error.

Equation	Coefficients	P	RSE
$T = a + bY$			
$T = a + bY + cY^2$			
$T = a + bY + cY^2 + dY^3$			
$T = a + bY + cY^2 + dY^3 + eY^4$			
$T = a + bY + cY^2 + dY^3 + eY^4 + fY^5$			
$T = a + bY + cY^2 + dY^3 + eY^4 + fY^5 + gY^6$			

b. After making this table, choose the polynomial of the function that you believe fits the data best. Provide reasoning and explanation why you chose the polynomial that you did.

c. Finally show the plot of your selected polynomial with the data.

Exercise 8.4. An equation that relates a consumer's nutrient content (denoted as y) to the nutrient content of food (denoted as x) is given by: $y = cx^{1/\theta}$, where $\theta \geq 1$ and $c > 0$ are both constants.

a. Use the dataset `phosphorous` to make a scatterplot with `algae` as the predictor (independent) variable and `daphnia` the response (dependent) variable.

b. Show that you can linearize the equation $y = cx^{1/\theta}$ with logarithms.

c. Determine a linear regression fit for your new linear equation.

d. Determine the value of c and θ in the original equation with the parameters from the linear fit.

Exercise 8.5. Similar to Exercise 8.4, do a non-linear least squares fit for the dataset `phosphorous` to the equation $y = cx^{1/\theta}$. For a starting point, you may use the values of c and θ from Exercise 8.4. Then make a plot of the original `phosphorous` data with the fitted model results.

Exercise 8.6. Example 8.1 guided you through the process to linearize the following equation:

$$V = \frac{V_{max}s}{s + K_m},$$
(8.5)

where s is the amount of substrate, K_m is half-saturation constant, and V_{max} the maximum reaction rate. (Typically V is used to signify the "velocity" of the reaction.) When doing a fit of the reciprocal of s with the reciprocal of V, what are the resulting values of V_{max} and K_m?

Exercise 8.7. Following from Example 8.1 and Exercise 8.6, apply the command nls to conduct a nonlinear least squares fit of the enzyme data to the equation:

$$V = \frac{V_{max}s}{s + K_m}, \qquad (8.6)$$

where s is the amount of substrate, K_m is the half-saturation constant, and V_{max} the maximum reaction rate. As starting points for the nonlinear least squares fit, you may use the values of K_m and V_{max} that were determined from Example 8.1. Then make a plot of the actual data with the fitted model curve.

Exercise 8.8. Consider the following data which represent the temperature over the course of a day:

Hour	Temperature
0	54
1	53
2	55
3	54
4	58
5	58
6	61
7	63
8	67
9	66
10	67
11	69
12	68
13	68
14	66
15	67
16	63
17	60
18	59
19	57
20	56
21	53
22	52
23	54
24	53

a. Make a scatterplot of these data, with the variable **Hour** on the horizontal axis.

b. A function that describes these data is $T = A + B\sin\left(\frac{\pi}{12} \cdot H\right) + C\cos\left(\frac{\pi}{12} \cdot H\right)$, where H is the hour and T is the temperature. Explain why this equation is linear for the parameters A, B, and C.

c. Define a `tibble` that includes the variables T, $\sin\left(\frac{\pi}{12} \cdot H\right)$, $\cos\left(\frac{\pi}{12} \cdot H\right)$.

d. Do a linear fit on your new data frame to report the values of A, B, and C.

e. Define a new `tibble` that has a sequence in H starting at 0 from 24 with at least 100 data points, and a value of T (`T_fitted`) using your coefficients of A, B, and C.

f. Add your fitted curve to the scatterplot. How do your fitted values compare to the data?

Exercise 8.9. Use the data from Exercise 8.8 to conduct a nonlinear fit (use the function `nls`) to the equation $T = A + B\sin\left(\frac{\pi}{12} \cdot H\right) + C\cos\left(\frac{\pi}{12} \cdot H\right)$. Good starting points are $A = 50$, $B = 1$, and $C = -10$.

9

Probability and Likelihood Functions

In Chapter 8 we began to the process of parameter estimation. We revisit parameter estimation here by applying *likelihood functions*, which is a topic from probability and statistics. Probability is the association of a set of observable events to a quantitative scale between 0 and 1. Informally, a value of zero means that event is not possible; 1 means that it definitely can happen.[1] We will only consider continuous events with the range of parameter estimation problems examined here.

This chapter will introduce likelihood functions but also discuss some interesting visualization techniques of multivariable functions and contour plots. As with Chapter 8 we are starting to build out some R skills and techniques that you can apply in other contexts. Let's get started!

9.1 Linear regression on a small dataset

Table 9.1 displays a dataset with a limited number of points where we wish to fit the function $y = bx$:

TABLE 9.1 A small, limited dataset.

x	1	2	4	4
y	3	5	4	10

For this example we are forcing the intercept term to equal zero - for most cases you will just fit the linear equation (see Exercise 9.6 where you will consider the intercept a). Figure 9.1 displays a quick scatterplot of these data:

---

[1]See Devore, Berk, and Carlton (2021) for a more refined definition of probability.

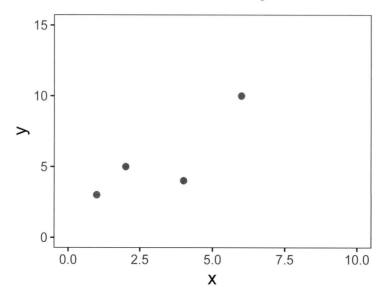

FIGURE 9.1 A scatterplot of a small, limited dataset (Table 9.1).

The goal here is to work to determine the value of b that is most *likely* - or consistent - with the data. However, before we tackle this further we need to understand how to quantify most *likely* in a mathematical sense. In order to do this, we need to take a quick excursion into continuous probability distributions.

9.2   Continuous probability density functions

Consider Figure 9.2, which may be familiar to you as the normal distribution or the bell curve:

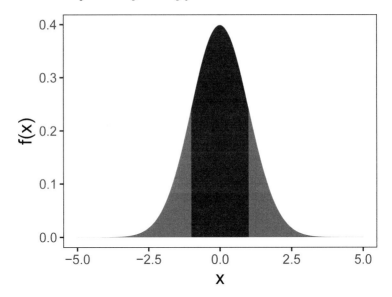

FIGURE 9.2 The standard normal distribution, with a shaded area between $x = \pm 1$

We tend to think of the plot and the associated function $f(x)$ as something with input and output (such as $f(0) = 0.3989$). However because it is a probability density function, the *area* between two points yields the probability of an event to fall within two values as shown in Figure 9.2.

In this case, the numerical value of the shaded area would represent the probability that our measurement is in the interval $-1 \leq x \leq 1$. The value of the area, or the probability, is 0.68269. Perhaps when you studied calculus the area was expressed as a definite integral: $\int_{-1}^{1} f(x) \, dx = 0.68269$, where $f(x)$ is the formula for the probability density function for the normal distribution. Here, the formula for the normal distribution $f(x)$ is given by Equation (9.1), where μ is the mean and σ is the standard deviation.[2]

$$f(x) = \frac{1}{\sqrt{2\pi}\sigma} e^{-(x-\mu)^2/(2\sigma^2)} \tag{9.1}$$

With this intuition we can summarize key facts about probability density functions:

---

[2]Were you ever asked to find the antiderivative of $\int e^{-x^2} \, dx$? You may recall that there is easy antiderivative - and to find values of definite integrals you need to approximate numerically. Thankfully R does that work using sophisticated numerical integration techniques!

- $f(x) \geq 0$ (this means that probability density functions are positive values)
- Area integrates to one (in probability, this means we have accounted for all of our outcomes)

9.3   Connecting probabilities to linear regression

Now that we have made that small excursion into probability, let's return to the parameter estimation problem. Another way to phrase this problem is to examine the probability distribution of the model-data residual for each measurement ϵ_i (Equation (9.2)):

$$\epsilon_i = y_i - f(x_i, \vec{\alpha}). \tag{9.2}$$

The approach with likelihood functions assumes a particular probability distribution on each residual. One common assumption is that the model-data residual is normally distributed. In most applications the mean of this distribution is zero ($\mu = 0$) and the standard deviation σ (which could be specified as measurement error, etc.). We formalize this assumption with a likelihood function L in Equation (9.3).

$$L(\epsilon_i) = \frac{1}{\sqrt{2\pi}\sigma} e^{-\epsilon_i^2 / 2\sigma^2} \tag{9.3}$$

To extend this further across all measurements, we use the idea of *independent, identically distributed* measurements so the joint likelihood of **all** the residuals (each ϵ_i) is the product of the individual likelihoods (Equation (9.4). The assumption of independent, identically distributed is a common one. As a note of caution you should always evaluate if this is a valid assumption for more advanced applications.

$$L(\vec{\epsilon}) = \prod_{i=1}^{N} \frac{1}{\sqrt{2\pi}\sigma} e^{-\epsilon_i^2 / 2\sigma^2} \tag{9.4}$$

We are making progress here; however to fully characterize the solution we need to specify the parameters $\vec{\alpha}$. A simple redefining of the likelihood function where we specify the measurements (x and y) and parameters ($\vec{\alpha}$) is all we need (Equation (9.5)).

$$L(\vec{\alpha}|\vec{x}, \vec{y}) = \prod_{i=1}^{N} \frac{1}{\sqrt{2\pi}\sigma} \exp(-(y_i - f(x_i, \vec{\alpha}))^2 / 2\sigma^2) \tag{9.5}$$

Now with Equation (9.5) we have a function where the best parameter estimate is the one that optimizes the likelihood.

Returning to our original linear regression problem (Table 9.1 and Figure 9.1), we want to determine the b for the function $y = bx$. Equation (9.6) then characterizes the likelihood of b, given the data $\vec{x}$ and $\vec{y}$:

$$L(b|\vec{x}, \vec{y}) = \left(\frac{1}{\sqrt{2\pi}\sigma}\right)^4 e^{-\frac{(3-b)^2}{2\sigma^2}} \cdot e^{-\frac{(5-2b)^2}{2\sigma^2}} \cdot e^{-\frac{(4-4b)^2}{2\sigma^2}} \cdot e^{-\frac{(10-4b)^2}{2\sigma^2}} \tag{9.6}$$

For the purposes of our argument here, we will assume $\sigma = 1$. Figure 9.3 shows a plot of the likelihood function $L(b|\vec{x}, \vec{y})$.

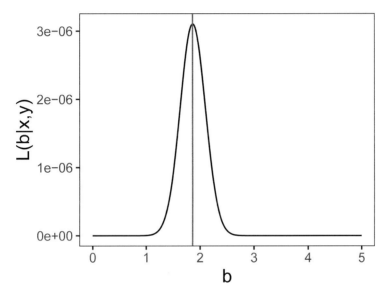

FIGURE 9.3 The likelihood function (Equation (9.6)) for the small dataset (Table 9.1), with the value of the maximum likelihood at $b = 1.865$ marked with a vertical line.

Note that in Figure 9.3 the values of $L(b|\vec{x}, \vec{y})$ on the vertical axis are really small. (This typically may be the case; see Exercise 9.2.) An alternative to the small numbers in $L(b)$ is to use the log-likelihood (Equation (9.7)):

$$\ln(L(\vec{\alpha}|\vec{x}, \vec{y})) = N \ln\left(\frac{1}{\sqrt{2\pi}\sigma}\right) - \sum_{i=1}^{N} \frac{(y_i - f(x_i, \vec{\alpha})^2}{2\sigma^2}$$

$$= -\frac{N}{2}\ln(2) - \frac{N}{2}\ln(\pi) - N\ln(\sigma) - \sum_{i=1}^{N} \frac{(y_i - f(x_i, \vec{\alpha})^2}{2\sigma^2} \tag{9.7}$$

In Exercise 9.5 you will be working on how to transform the likelihood function $L(b)$ to the log-likelihood $\ln(L(b))$ and showing that Equation (9.6) is maximized at $b = 1.865$. The data with the fitted line is shown in Figure 9.4.

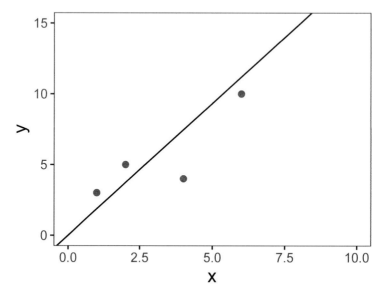

FIGURE 9.4 A scatterplot of a small dataset (Table 9.1) with fitted line $y = 1.865x$ from optimizing Equation (9.6).

9.4   Visualizing likelihood surfaces

Next we are going to examine a second example from Gause (1932) which modeled the growing of yeast in solution. This classic paper examines the biological principal of *competitive exclusion,* how one species can out-compete another one for resources. Some of the data from Gause (1932) is encoded in the data frame yeast in the demodelr package. For this example we are going to examine a model for one species growing without competition. Figure 9.5 shows a scatterplot of the yeast data.

```
### Make a quick ggplot of the data

ggplot() +
  geom_point(
    data = yeast,
    aes(x = time, y = volume),
    color = "red",
    size = 2
```

```
) +
labs(x = "Time", y = "Volume")
```

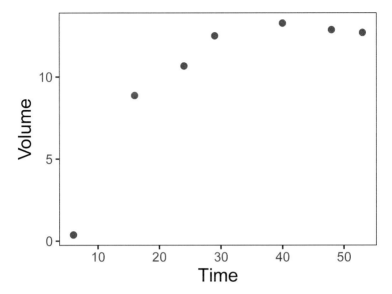

FIGURE 9.5 Scatterplot of *Sacchromyces* volume growing by itself in a container.

We are going to assume the population of yeast (represented with the measurement of volume) over time changes according to the differential equation $\frac{dy}{dt} = -by\frac{(K-y)}{K}$, where y is the population of the yeast, and b represents the growth rate, and K is the carrying capacity of the population.

Equation (9.8) shows the solution to this differential equation, where the additional parameter a can be found through application of the initial condition y_0.

$$y = \frac{K}{1 + e^{a-bt}} \tag{9.8}$$

In Gause (1932) the value of a was determined by solving the initial value problem $y(0) = 0.45$. In Exercise 9.1 you will show that $a = \ln\left(\frac{K}{0.45} - 1\right)$.

Equation (9.8) then has two parameters: K and b. Here we are going to explore the likelihood function to try to determine the best set of values for the two parameters K and b using a function in the demodelr package called compute_likelihood. Inputs to the compute_likelihood function are the following:

- A function $y = f(x, \vec{\alpha})$
- A dataset $(\vec{x}, \vec{y})$
- Ranges of your parameters $\vec{\alpha}$.

The `compute_likelihood` function also has an optional input `logLikely` that allows you to specify if you want to compute the likelihood or the log-likelihood. The default is that `logLikely` is `FALSE`, meaning that the normal likelihoods are plotted.

First we will define the equation used to compute our model in the likelihood. As with the functions `euler` or `systems` in Chapter 4 we need to define this function:

```
library(demodelr)

# Define the solution to the differential equation with
# parameters K and bGause model equation
gause_model <- volume ~ K / (1 + exp(log(K / 0.45 - 1) - b * time))


# Identify the ranges of the parameters that we wish to investigate
kParam <- seq(5, 20, length.out = 100)
bParam <- seq(0, 1, length.out = 100)


# Allow for all the possible combinations of parameters
gause_parameters <- expand.grid(K = kParam, b = bParam)

# Now compute the likelihood
gause_likelihood <- compute_likelihood(
  model = gause_model,
  data = yeast,
  parameters = gause_parameters,
  logLikely = FALSE
  )
```

Ok, let's break this code down step by step:

- The line `gause_model <- volume ~ K/(1+exp(log(k/0.45-1)-b*time))` identifies the formula that relates the variables `time` to `volume` in the dataset `yeast`.
- We define the ranges (minimum and maximum values) for our parameters by defining a sequence. Because we want to look at *all possible combinations* of these parameters we use the command `expand.grid`.
- The input `logLikely = FALSE` to `compute_likelihood` reports back likelihood values.

Some care is needed in defining the number of points (`length.out = 100`) in the sequence that we want to evaluate - we will have 100^2 different combinations of K and b on our grid, which does take time to evaluate.

The output to `compute_likelihood` is a list, which is a flexible data structure in R. You can think of this as a collection of items - which could be data frames of different sizes. In this case, what gets returned are two data frames: `likelihood`, which is a data frame of likelihood values for each of the parameters and `opt_value`, which reports back the values of the parameters that optimize the likelihood function. Note that the optimum value is *an approximation*, as it is just the optimum from the input values of K and b provided on our grid. Let's take a look at the reported optimum values, which we can do with the syntax `LIST_NAME$VARIABLE_NAME`, where the dollar sign ($) helps identify which variable from the list you are investigating.

```
gause_likelihood$opt_value
```

```
## # A tibble: 1 x 4
##       K      b  l_hood log_lik
##   <dbl> <dbl>   <dbl> <lgl>
## 1  12.7 0.242 0.000348 FALSE
```

It is also important to visualize this likelihood function. For this dataset we have the two parameters K and b, so the likelihood function will be a *likelihood surface*, rather than a two-dimensional plot. To visualize this in R we can use a contour diagram. Figure 9.6 displays this countour plot.

```
# Define the likelihood values
my_likelihood <- gause_likelihood$likelihood

# Make a contour plot
ggplot(data = my_likelihood) +
  geom_tile(aes(x = K, y = b, fill = l_hood)) +
  stat_contour(aes(x = K, y = b, z = l_hood))
```

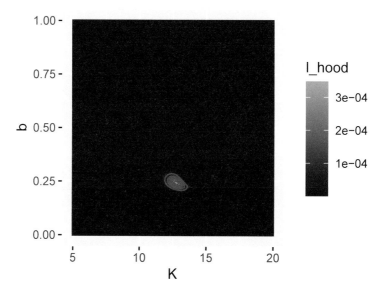

FIGURE 9.6 Likelihood surface and contour lines for the yeast dataset.

Similar to before, let's take this step by step:

- The command my_likelihood just puts the likelihood values in a data frame.
- The ggplot command is similar to that used before.
- We use geom_tile to visualize the likelihood surface. There are three required inputs from the my_likelihood data frame: the x and y axis data values and the fill value, which represents the height of the likelihood function.
- The command stat_contour draws the contour lines, or places where the likelihood function has the same value. Notice how we used z = l_hood rather than fill here. This function helps "smooth" out any jaggedness in the contours.

In Figure 9.6 there appears to be a large region where the likelihood has the same value. (Admittedly I chose some broad parameter ranges for K and b). We can refine that by producing a second contour plot that focuses in on parameters closer to the calculated optimum value at $K = 13$ and $b = 0.07$ (Figure 9.7):

```
## # A tibble: 1 x 4
##       K     b   l_hood log_lik
##    <dbl> <dbl>    <dbl> <lgl>
## 1  12.8 0.241 0.000349 FALSE
```

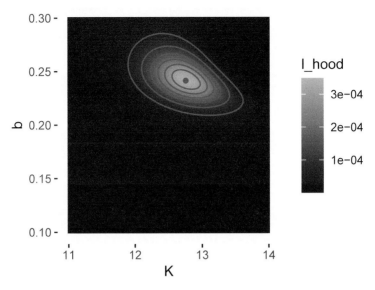

FIGURE 9.7 Zoomed in likelihood surface. for the yeast dataset. The computed location of the optimum value is shown as a red point.

The reported values for K (12.8) and b (0.241) may be close to what was reported from Figure 9.6. Notice that in Figure 9.7 I also added in the location of the optimum point with geom_point().

As a final step, once you have settled on the value that optimizes the likelihood function, is to compare the optimized parameters against the data (Figure 9.8):

```
# Define the parameters and the times to evaluate:
my_params <- gause_likelihood_rev$opt_value
time <- seq(0, 60, length.out = 100)

# Get the right hand side of your equations
new_eq <- gause_model %>%
  formula.tools::rhs()

# This collects the parameters and data into a list
in_list <- c(my_params, time) %>% as.list()

# The eval command evaluates your model
out_model <- eval(new_eq, envir = in_list)


# Now collect everything into a data frame:
my_prediction <- tibble(time = time, volume = out_model)
```

```
ggplot() +
  geom_point(
    data = yeast,
    aes(x = time, y = volume),
    color = "red",
    size = 2
  ) +
  geom_line(
    data = my_prediction,
    aes(x = time, y = volume)
  ) +
  labs(x = "Time", y = "Volume")
```

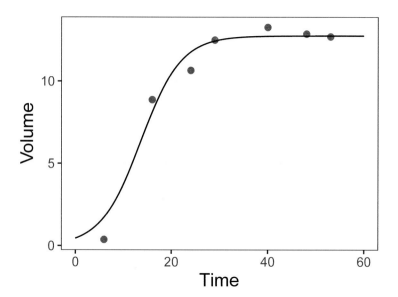

FIGURE 9.8 Model and data comparison of the `yeast` dataset from maximum likelihood estimation.

All right, this code block has some new commands and techniques that need explaining. Once we have the parameter estimates we need to compute the modeled values.

- First we define the `params` and the `time` we wish to evaluate with our model.
- We need to evaluate the right hand side of $y = \dfrac{K}{1 + e^{a+bt}}$, so the definition of `new_eq` helps to do that, using the package `formula.tools`.

- The %>% is the tidyverse pipe[3]. This is a very useful command to help make code more readable!
- in_list <- c(params,my_time) %>% as.list() collects the parameters and input times in one list to evaluate the model with out_model <- eval(new_eq,envir=in_list)
- We make a data frame called my_prediction so we can then plot.

And the rest of the plotting commands you should be used to. This activity focused on the likelihood function - Exercise 9.3 has you repeat this analysis with the log-likelihood function.

9.5 Looking back and forward

This chapter covered a lot of ground - from probability and likelihood functions to computing and visualizing these. A good strategy for a likelihood function is to visualize the function and then explore to find values that optimize the likelihood or log-likelihood function. This approach is one example of *successive approximations* or using an iterative method to determine a solution. While this chapter focused on optimizing a likelihood function with one or two parameters, the successive approximation method does generalize to more parameters. However searching for parameters becomes tricky (read: tedious and slow) in high-dimensional spaces. In later chapters we will explore numerical methods to accelerate convergence to an optimum value.

9.6 Exercises

Exercise 9.1. Algebraically solve the equation $0.45 = \dfrac{K}{1 + e^a}$ for a.

Exercise 9.2. Compute the values of $L(b|\vec{x}, \vec{y})$ and $\ln(L(b|\vec{x}, \vec{y}))$ for each of the data points in Equation (9.6) when $b = 1.865$ and $\sigma = 1$. (This means that these 4 values would be multiplied or added when you compute the full likelihood function.) Explain if the likelihood or log-likelihood would be easier to calculate in instances when the number of observations is large.

Exercise 9.3. Visualize the likelihood function for the yeast dataset, but in this case report out and visualize the log-likelihood. (This means that you are setting the option logLikely = TRUE in the compute_likelihood function.) Compare the log-likelihood surface to Figure 9.7.

---

[3]https://r4ds.had.co.nz/pipes.html#pipes

Exercise 9.4. When we generated our plot of the likelihood function in Figure 9.3 we assumed that $\sigma = 1$ in Equation (9.6). For this exercise you will explore what happens in Equation (9.6) as σ increases or decreases.

a. Use desmos (www.desmos.com/calculator) or R to generate a plot of Equation (9.6). What happens to the shape of the likelihood function as σ increases?
b. How does the estimate of b change as σ changes?
c. The spread of the distribution (in terms of it being more peaked or less peaked) is a measure of uncertainty of a parameter estimate. How does the resulting parameter uncertainty change as σ changes?

Exercise 9.5. Using Equation (9.6) with $\sigma = 1$:

a. Apply the natural logarithm to both sides of this expression. Using properties of logarithms, show that the log-likelihood function is $\ln(L(b)) =$
$$-2\ln(2) - 2\ln(\pi) - \frac{(3-b)^2}{2} - \frac{(5-2b)^2}{2} - \frac{(4-4b)^2}{2} - \frac{(10-4b)^2}{2}.$$
b. Make a plot of the log-likelihood function (in desmos or R).

c. At what values of b Where is this function optimized? Does your graph indicate that it is a maximum or a minimum value?

d. Compare this likelihood estimate for b to what was found in Figure 9.3.

Exercise 9.6. Consider the linear model $y = a + bx$ for the following dataset:

x	y
1	3
2	5
4	4
4	10

a. With the function compute_likelihood, generate a contour plot of both the likelihood and log-likelihood functions. You may assume $0 \le a \le 5$ and $0 \le b \le 5$.
b. Make a scatterplot of these data with the equation $y = a + bx$ with your maximum likelihood parameter estimates.
c. Earlier when we fit $y = bx$ we found $b = 1.865$. How does adding a as a model parameter affect your estimate of b?

Exercise 9.7. For the function $P(t) = \dfrac{K}{1 + e^{a+bt}}$, with $P(0) = P_0$, determine an expression for the parameter a in terms of K, b, and P_0.

Exercise 9.8. The values returned by the maximum likelihood estimate for Equation (9.8) were a little different from those reported in Gause (1932):

Parameter	Maximum Likelihood Estimate	Gause (1932)
K	12.7	13.0
b	0.24242	0.21827

Using the yeast dataset, plot the function $y = \dfrac{K}{1 + e^{a-bt}}$ (setting $a = \ln\left(\dfrac{K}{0.45} - 1\right)$) using both sets of parameters. Which approach (the Maximum Likelihood estimate or Gause (1932)) does a better job representing the data?

Exercise 9.9. An equation that relates a consumer's nutrient content (denoted as y) to the nutrient content of food (denoted as x) is given by: $y = cx^{1/\theta}$, where $\theta \geq 1$ and $c > 0$ are both constants.

a. Use the dataset phosphorous to make a scatterplot with the variable algae on the horizontal axis, daphnia on the vertical axis.
b. Generate a contour plot for the likelihood function for these data. You may assume $0 \leq c \leq 5$ and $1 \leq \theta \leq 20$. What are the values of c and θ that optimize the likelihood? *Hint:* for the dataset phosphorous be sure to use the variables $x =$algae and $y =$daphnia.
c. Add the fitted curve to your scatterplot and evaluate your fitted results.

Exercise 9.10. A dog's weight W (pounds) changes over D days according to the following function:

$$W = f(D, p_1, p_2) = \frac{p_1}{1 + e^{2.462 - p_2 D}}, \qquad (9.9)$$

where p_1 and p_2 are parameters.

a. This function can be used to describe the data wilson. Make a scatterplot with the wilson data. What is the long term weight of the dog?
b. Generate a contour plot for the likelihood function for these data. What are the values of p_1 and p_2 that optimize the likelihood? *You may assume that p_1 and p_2 are both positive.*
c. With your values of p_1 and p_2 add the function W to your scatterplot and compare the fitted curve to the data.

10

Cost Functions and Bayes' Rule

Chapter 9 introduced likelihood functions as an approach to tackle parameter estimation. However this is not the only approach to understand model-data fusion. This chapter introduces *cost functions*, which estimates parameters from data using a least squares approach. Want to know a secret? Cost functions are very closely related to log-likelihood functions. This chapter will explore this idea some more, first by exploring model-data residuals, defining a cost function, and then connecting them back to likelihood functions. To complete the circle, this chapter ends by discussing Bayes' Rule, which will further strengthen the connection between cost and likelihood functions. Let's get started!

10.1 Cost functions and model-data residuals

Let's revisit the linear regression problem from Chapter 9. Recall Table 9.1 from Chapter 9. With these data we wanted to fit a function of the form $y = bx$ (forcing the intercept term to be zero). We will extend Table 9.1 to include the model-data residual computed as $y - bx$ in Table 10.1:

TABLE 10.1 A small, limited dataset (Table 9.1) with the computed model-data residual with parameter b, along with model-data residuals for different values of b.

x	y	bx	$y - bx$	$b = 1$	$b = 3$	$b = -1$
1	3	b	$3 - b$	2	0	4
2	5	$2b$	$5 - 2b$	3	-1	7
4	4	$4b$	$4 - 4b$	0	-8	8
4	10	$4b$	$10 - 4b$	6	-2	14

Also included in Table 10.1 are the model-data residual values for different values of b. Notably values of the residuals can be negative and some can be positive - which makes it tricky to assess the "best" value of b from the residuals alone. (If we found a value of b where the residuals were all zero, then we would have the "best" value of b![1]).

To assess the overall residuals as a function of the value of b, we need to take into consideration not just the value of the residual (positive or negative), but rather some way to measure the overall distance of *all* the residuals from a given value of b. One way to define that is with a function that squares each residual (so that negative and positive values don't cancel each other) and adds each of those results together. We call this the *sum squared residuals*. So for example, the sum squared residual when $b = 1$ is shown in Equation (10.1):

$$\text{Sum square residual: } 2^2 + 3^2 + 0^2 + 6^2 = 49 \qquad (10.1)$$

The other square residuals are 68 when $b = 3$ and 325 when $b = -1$. So of these choices for b, the one that minimizes the square residual is $b = 1$.

Let's generalize this to determine a function to compute the sum square residual for any value of b. This function, denoted as $S(b)$, is called the cost function (Equation (10.2)):

$$S(b) = (3 - b)^2 + (5 - 2b)^2 + (4 - 4b)^2 + (10 - 4b)^2 \qquad (10.2)$$

Equation (10.2) is a function of one variable (b). Figure 10.1 shows a graph of $S(b)$. Notice how the plot of $S(b)$ is a nice quadratic function, with a minimum at $b = 1.865$. Did you notice that this value for b is the same value for the minimum that we found from Equation (9.6) in Chapter 9? In Exercise 10.1 you will use calculus to determine the optimum value of $S(b)$.

---

[1] To be fair, that means the data would be perfectly on a line; not too interesting of a problem, right?

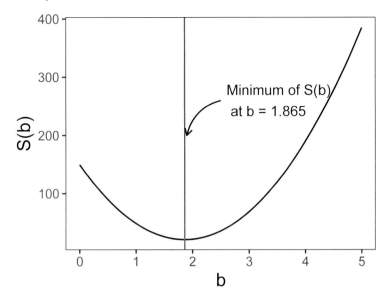

FIGURE 10.1 Plot of Equation (10.2). The vertical line denotes the minimum value at $b = 1.865$.

10.1.1 Accounting for uncertainty

The cost function can also incorporate uncertainty in the value of the response variable y. We will define this uncertainty as σ and have it be the same for each value y_i. In some cases the uncertainty may vary from measurement to measurement - but the concepts presented here can generalize. To account for this uncertainty we divide each of the square residuals in Equation (10.2) by σ^2, as shown in Equation (10.3) using $\sum$ notation.

$$S(\vec{\alpha}) = \sum_{i=1}^{N} \frac{(y_i - f(x, \vec{\alpha}))^2}{\sigma^2} \tag{10.3}$$

As an example, comparing Equation (10.2) to Equation (10.3) we have $N = 4$, $f(x_i, \vec{\alpha}) = bx$, and $\sigma = 1$.

10.1.2 Comparing cost and log-likelihood functions

Chapter 9 defined the log-likelihood function (Equation (9.7)), which for the small dataset we are studying is represented with Equation (10.4) where $\sigma = 1$ and $N = 4$:

Probability	Optimistic	Pessimistic	Total
Voted for the incumbent President	0.20	0.20	0.40
Did not vote for incumbent President	0.15	0.45	0.60
Total	0.35	0.65	1.00

Compute the probability of having an optimistic view on the economy.

Solution. Based on the probability table, we define the following probabilities:

- The probability you voted for the incumbent President *and* have an **optimistic** view on the economy is 0.20
- The probability you **did not** vote for the incumbent President *and* have an **optimistic** view on the economy is 0.15
- The probability you voted for the incumbent President *and* have an **pessimistic** view on the economy is 0.20
- The probability you **did not** vote for the incumbent President *and* have an **pessimistic** view on the economy is 0.45

We calculate the probability of having an **optimistic view on the economy** by adding the probabilities with an optimistic view, whether or not they voted for the incumbent President. For this example, this probability sums to $0.20 + 0.15 = 0.35$, or 35%.

On the other hand, the probability you have a pessimistic view on the economy is $0.20 + 0.45 = 0.65$, or 65%. Notice how the two of these together (probability of optimistic and pessimistic views of the economy is 1, or 100% of the outcomes.)

10.3.1   Conditional probabilities

Next, let's discuss conditional probabilities. A conditional probability is the probability of an outcome given some previous outcome, or $\Pr(A|B)$, where $\Pr$ means "probability of an outcome" and A and B are two different outcomes or events. In probability theory you might study the following law of conditional probability:

$$
\begin{aligned}
\Pr(A \text{ and } B) &= \Pr(A \text{ given } B) \cdot \Pr(B) \\
&= \Pr(A|B) \cdot \Pr(B) \\
&= \Pr(B|A) \cdot \Pr(A)
\end{aligned}
\tag{10.6}
$$

Typically when expressing conditional probabilities we remove "and" and write $P(A \text{ and } B)$ as $P(AB)$ and "given" as $P(A \text{ given } B)$ as $P(A|B)$.

Example 10.2. Continuing with Example 10.1, sometimes people believe that your views of the economy influence whether you are going to vote for the incumbent President in an election.[2] Use the information from the table in Example 10.1 to compute the probability you voted for the incumbent President *given* you have an optimistic view of the economy.

Solution. To compute the probability you voted for the incumbent President *given* you have an optimistic view of the economy is a rearrangement of Equation (10.6):

$$\text{Pr(Voted for incumbent President | Optimistic View on Economy)} =$$

$$\frac{\text{Pr(Voted for incumbent President and Optimistic View on Economy)}}{\text{Pr(Optimistic View on Economy)}} =$$

$$\frac{0.20}{0.35} = 0.57$$
(10.7)

So if you have an optimistic view on the economy, there is a 57% chance you will vote for the incumbent President. Contrast this result to the probability that you voted for the incumbent President (Example 10.1), which is only 40%. Perhaps your view of the economy does indeed influence whether or not you would vote to re-elect the incumbent President.

10.3.2 Bayes' rule

Using the incumbent President and economy example as a framework, we will introduce *Bayes' Rule*[3], which is a re-arrangment of the rule for conditional probability:

$$\text{Pr}(A|B) = \frac{\text{Pr}(B|A) \cdot \text{Pr}(A)}{\text{Pr}(B)}$$
(10.8)

It turns out Bayes' Rule is a really helpful way to understand how we can systematically incorporate this prior information into the likelihood function (and by association the cost function). For parameter estimation our goal is to estimate parameters, given the data. Another way to state Bayes' Rule in Equation (10.8) is using terms of parameters and data:

$$\text{Pr(parameters | data)} = \frac{\text{Pr(data | parameters)} \cdot \text{Pr(parameters)}}{\text{Pr(data)}}$$
(10.9)

---

[2]https://www.cbsnews.com/news/how-much-impact-can-a-president-have-on-the-economy/
[3]https://en.wikipedia.org/wiki/Bayes%27_theorem

While Equation (10.9) seems pretty conceptual, here are some key highlights:

- In practice, the term Pr(data | parameters) in Equation (10.9) is the likelihood function (Equation (9.5)).
- The term Pr(parameters) is the probability distribution of the *prior information* on the parameters, specifying the probability distribution functions for the given context. When this distribution is the same as Pr(data | parameters) (typically normally distributed), prior information has a multiplicative effect on the likelihood function (Pr(parameters | data)). (Or an additive effect on the log-likelihood function.) *This is good news!* When we added that additional term for prior information into $\tilde{S}(b)$ in Equation (10.5), we accounted for the prior information correctly. In Exercise 10.6 you will explore how the log-likelihood is related to the cost function.
- The expression Pr(parameters | data) is the start of a framework for a probability density function, which should integrate to unity. (You will explore this more if you study probability theory.) This denominator term is called a normalizing constant[4]. Since our overall goal is to select parameters that optimize Pr(parameters | data), the expression in the denominator (Pr(data)) does not change the *location* of the optimum values.

10.4   Bayes' rule in action

Wow - we made some significant progress in our conceptual understanding of how to incorporate models and data! Let's see how this applies to our linear regression problem $(y = bx)$. We have the following assumptions:

- **Assumption 1:** The data are independent, identically distributed. We can then write the likelihood function as the following:

$$\Pr(\vec{y}|b) = \left(\frac{1}{\sqrt{2\pi}\sigma}\right)^4 e^{-\frac{(3-b)^2}{2\sigma^2}} \cdot e^{-\frac{(5-2b)^2}{2\sigma^2}} \cdot e^{-\frac{(4-4b)^2}{2\sigma^2}} \cdot e^{-\frac{(10-4b)^2}{2\sigma^2}} \qquad (10.10)$$

- **Assumption 2:** Prior knowledge expects us to say that b is normally distributed with mean 1.3 and standard deviation 0.1. Incorporating this information allows us to write the following:

$$\Pr(b) = \frac{1}{\sqrt{2\pi} \cdot 0.1} e^{-\frac{(b-1.3)^2}{2 \cdot 0.1^2}} \qquad (10.11)$$

When we combine the two pieces of information, the probability of b, given the data $\vec{y}$, is the following:

---

[4]https://stats.stackexchange.com/questions/12112/normalizing-constant-in-bayes-theorem

$$\Pr(b|\vec{y}) \approx e^{-\frac{(3-b)^2}{2\sigma^2}} \cdot e^{-\frac{(5-2b)^2}{2\sigma^2}} \cdot e^{-\frac{(4-4b)^2}{2\sigma^2}} \cdot e^{-\frac{(10-4b)^2}{2\sigma^2}} \cdot e^{-\frac{(b-1.3)^2}{2\cdot0.1^2}} \quad (10.12)$$

Notice we are ignoring the terms $\left(\dfrac{1}{\sqrt{2\pi}\cdot\sigma}\right)^4$ and $\dfrac{1}{\sqrt{2\pi}\cdot0.1}$, because per our discussion above not including them does not change the *location* of the optimum value, only the value of the likelihood function. The plot of $\Pr(b|\vec{y})$, assuming $\sigma=1$ is shown in Figure 10.3:

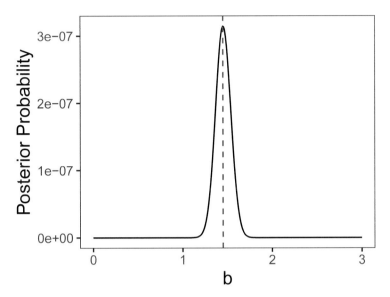

FIGURE 10.3 Equation (10.12) with optimum value at $b=1.45$ denoted in a blue dashed line.

It looks like the value that optimizes our posterior probability is $b=1.45$. This is similar the value of $\tilde{b}$ from Equation (10.5). Again, *this is no coincidence*. Adding in prior information to the cost function or using Bayes' Rule are equivalent approaches.

10.5   Next steps

Now that we have seen the usefulness of cost functions and Bayes' Rule we can begin to apply this to larger problems involving more equations and data. In order to do that we need to explore some computational methods to scale this problem up - which we will do in subsequent chapters.

10.6   Exercises

Exercise 10.1. The following problem works with Table 9.1 to determine the value of b with the function $y = bx$ as in this chapter.

a. Using calculus, show that the cost function $S(b) = (3 - b)^2 + (5 - 2b)^2 + (4 - 4b)^2 + (10 - 4b)^2$ has a minimum value at $b = 1.86$.
b. What is the value of $S(1.865)$?
c. Use a similar approach to determine the minimum of the revised cost function $\tilde{S}(b) = (3 - b)^2 + (5 - 2b)^2 + (4 - 4b)^2 + (10 - 4b)^2 + (b - 1.3)^2$. Call this value $\tilde{b}$.
d. How do the values of $S(1.865)$ and $\tilde{S}(\tilde{b})$ compare?
e. Make a plot of the cost functions $S(b)$ and $\tilde{S}(b)$ to verify the optimum values.
f. Make a scatter plot with the data and the function $y = bx$ and $y = \tilde{b}x$. How do the two estimates compare with the data?

Exercise 10.2. Use calculus to determine the optimum value of b for Equation (10.4). Do you obtain the same value of b for Equation (10.2)?

Exercise 10.3. (Inspired from Hugo van den Berg (2011)) Consider the nutrient equation $y = cx^{1/\theta}$ using the dataset `phosphorous`.

a. Write down a formula for the objective function $S(c, \theta)$ that characterizes this equation (that includes the dataset `phosphorous`).
b. Fix $c = 1.737$. Make a `ggplot` of $S(1.737, \theta)$ for $1 \leq \theta \leq 10$.
c. How many critical points does this function have over this interval? Which value of θ is the global minimum?

Exercise 10.4. Use the cost function $S(1.737, \theta)$ from Exercise 10.3 to answer the following questions:

a. Researchers believe that $\theta \approx 7$. Re-write $S(1.737, \theta)$ to account for this additional (prior) information.
b. How does the inclusion of this additional information change the shape of the cost function and the location of the global minimum?
c. Finally, reconsider the fact that $\theta \approx 7 \pm .5$ (as prior information). How does that modify $S(1.737, \theta)$ further and the location of the global minimum?

Exercise 10.5. One way to generalize the notion of prior information using cost functions is to include a term that represents the degree of uncertainty in the prior information, such as σ. For the problem $y = bx$ this leads to the following cost function: $\tilde{S}_{revised}(b) = (3 - b)^2 + (5 - 2b)^2 + (4 - 4b)^2 + (10 - 4b)^2 + \dfrac{(b - 1.3)^2}{\sigma^2}$.

Use calculus to determine the optimum value for $\tilde{S}_{revised}(b)$, expressed in terms of $\tilde{b}_{revised} = f(\sigma)$ (your optimum value will be a function of σ). What happens to $\tilde{b}_{revised}$ as $\sigma \to \infty$?

Exercise 10.6. For this problem you will minimize some generic functions.

a. Using calculus, verify that the optimum value of $y = ax^2 + bx + c$ occurs at $x = -\dfrac{b}{2a}$. (You can assume $a > 0$.)

b. Using calculus, verify that a critical point of $z = e^{-(ax^2+bx+c)^2}$ also occurs at $x = -\dfrac{b}{2a}$. Note: this is a good exercise to practice your differentiation skills!

c. Algebraically show that $\ln(z) = -y$.

d. Explain why y is similar to a cost function $S(b)$ and z is similar to a likelihood function.

Exercise 10.7. This problem continues the re-election of the incumbent President and viewpoint on the economy in Example 10.1. Determine the following conditional probabilities:

a. Determine the probability that you **voted for the incumbent President** given that you have a **pessimistic view on the economy**.

b. Determine the probability that you **did not vote for the incumbent President** given that you have an **pessimistic view on the economy**.

c. Determine the probability that you **did not vote for the incumbent President** given that you have an **optimistic view on the economy**.

d. Determine the probability that you have an **pessimistic view on the economy** given that you **voted for the incumbent President**.

e. Determine the probability that you have an **optimistic view on the economy** given that you **did not vote for the incumbent President**.

Exercise 10.8. Incumbents have an advantage in re-election due to wider name recognition, which may boost their re-election chances, as shown in the following table:

Probability	Being elected	Not being elected	Total
Having name recognition	0.55	0.25	0.80
Not having name recognition	0.05	0.15	0.20
Total	0.60	0.40	1.00

TABLE 11.1 Weather station data from a Minnesota snowstorm.

date	time	station_id	station_name	snowfall
4/16/18	5:00 AM	MN-HN-78	Richfield 1.9 WNW	22.0
4/16/18	7:00 AM	MN-HN-9	Minneapolis 3.0 NNW	19.0
4/16/18	7:00 AM	MN-HN-14	Minnetrista 1.5 SSE	12.5
4/16/18	7:00 AM	MN-HN-30	Plymouth 2.4 ENE	18.5
4/16/18	7:00 AM	MN-HN-58	Champlin 1.5 ESE (118)	20.0

An alternative approach to optimization relies on the idea of *sampling*, which randomly selects parameter values and then computes the cost function. After a certain amount of time or iterations, all the values of the cost function are compared to each other to determine the optimum value. We will refine the concept of sampling to determine the optimum value in subsequent chapters (Chapters 12 and 13), but this chapter develops some foundations in sampling - we will study how to plot histograms in R and then apply these to the *bootstrap* method, which relies on random sampling. Let's get started!

11.1   Histograms and their visualization

To introduce the idea of a histogram, consider Table 11.1, which is a sample of the dataset `snowfall` in the `demodelr` package. The `snowfall` dataset is measurements of precipitations collected at weather stations in the Twin Cities (Minneapolis, Saint Paul, and surrounding suburbs) from a spring snowstorm. These data come from a cooperative network of local observers[1]. Yes, it can snow in Minnesota in April. Sigh.

The header row in Table 11.1 also lists the names of the associated variables in this data frame. We are going to focus on the variable `snowfall`. A histogram is an easy way to view the distribution of measurements, using `geom_histogram` (Figure 11.2). You may recall that a histogram represents the frequency count of a random variable, typically binned over certain intervals.

```
ggplot(data = snowfall) +
  geom_histogram(aes(x = snowfall)) +
  labs(
    x = "Snowfall amount",
    y = "Number of observations"
  )
```

---

[1] https://www.cocorahs.org/ViewData/ListDailyPrecipReports.aspx

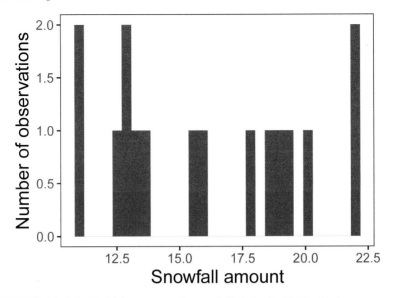

FIGURE 11.2 Initial histogram of snowfall data in Table 11.1.

This code introduces `geom_histogram`. To create the historgram we use the aesthetic (`aes(x = snowfall)`) to display the histogram for the `snowfall` column in the dataset `snowfall`.

At the R console you may have received a warning about such as `stat_bin()` using bins = 30. Pick better value with binwidth. The function `geom_histogram` doesn't guess a bin width, but one rule of thumb is to select the number of bins to be equal to the square root of the number of observations (16 in this instance). So let's adjust the number of bins to 4 in Figure 11.3.

```
ggplot() +
  geom_histogram(data = snowfall, aes(x = snowfall), bins = 4) +
  labs(
    x = "Snowfall amount",
    y = "Number of observations"
  )
```

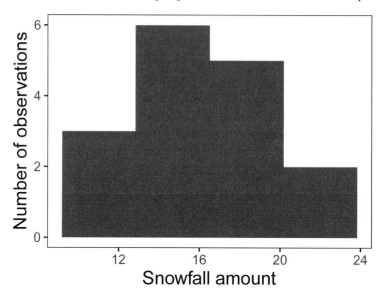

FIGURE 11.3 Adjusted histogram of snowfall data in Table 11.1 with the number of bins set to 4.

11.2   Statistical theory: sampling distributions

11.2.1   Sampling the empirical distribution

The histogram of snowfall measurements (Figure 11.3) from the snowfall data (Table 11.1) represents an *empirical probability distribution* of measurements. While it is great to have all these observations, a key question is: *What is an estimate for the representative amount of snowfall for this storm?* The snowfall measurements illustrate the difference between what statisticians call a *population* (the true distribution of measurements) and a *sample* (what people observe). To go a little deeper, let's define a random variable S for this population, which has an (unknown) probability density function associated with it. The measurements shown Table 11.1 are *samples* of this distribution. While the average of the precipitation data (16.1 inches) might be a defensible value for the average amount of snow that fell, what we don't know is *how well* this empirical mean approximates the expected value of the distribution for S.

Each of the entries in Table 11.1 represents a measurement made by a particular observer. To get the true distribution we would need to add more observers, but that isn't realistic for this case as the snowfall event has already passed - we can't go "back in time."

One way is to generate a *bootstrap sample*, which is a sample of the original dataset with replacement. The workflow that we will apply is the following:

Do once → Do several times → Visualize → Summarize

"Do once" is the first step, using sampling with replacement. This process is easily done with the R command `slice_sample` (you should try this out yourself), which in the following code assigns a sample to the variable `p_new`:

```
p_new <- slice_sample(snowfall, prop = 1, replace = TRUE)
```

Let's break this code down:

- We are sampling the `snowfall` data frame with replacement (`replace = TRUE`). If you are uncertain about how sampling with replacement works, here is one visual: say you have each of the snowfall measurements written on a piece of paper in a hat. You draw one slip of paper, record the measurement, and then place that slip of paper back in the hat to draw again, until you have as many measurements as the original data frame.

- The command `prop=1` means that we are sampling 100% of the `snowfall` data frame.

Once we have taken a sample, we can then compute the mean (average) and the standard deviation of the sample:

```
slice_sample(snowfall, prop = 1, replace = TRUE) %>%
  summarize(
    mean = mean(snowfall, na.rm = TRUE)
  )
```

```
##       mean
## 1 17.84375
```

How the above code works is to first do the sampling, and then the summary. The command `summarize` collapses the `snowfall` data frame and computes the mean and the standard deviation `sd` of the column `snowfall`. We have the command `na.rm=TRUE` to remove any `NA` values that may affect the computation.

"Do several times" is the second step. Here we are going to rely on some powerful functionality from the `purrr` package[2]. This package has the wonderful command `map_df`, which allows you to efficiently repeat a process several times and return a data frame as output. Evaluate the following code on your own:

---

[2]`https://purrr.tidyverse.org/`

```
purrr::map_df(
  1:10,
  ~ (
    slice_sample(snowfall, prop = 1, replace = TRUE) %>%
      summarize(
        mean = mean(snowfall, na.rm = TRUE)
      )
  ) # Close off the tilde ~ ()
) # Close off the map_df
```

Let's review this code step by step:

- The basic structure is `map_df(1:N,~(COMMANDS))`, where `N` is the number of times you want to run your code (in this case `N=10`).
- The second part `~(COMMANDS)` lists the different commands we want to re-run (here the resampling of our dataset and then subsequent summarizing).

What should be returned is a data frame that lists the `mean` of each bootstrap sample. The process of randomly sampling a dataset is called *bootstrapping*.

I can appreciate that this programming might be a little tricky to understand and follow - don't worry - the goal is to give you a tool that you can easily adapt to a situation.

"Visualize" is the step where we will use a histogram to examine the distribution of the bootstrap samples. The following code does this all, changing the number of bootstrap samples to 1000:

```
bootstrap_samples <- purrr::map_df(
  1:1000,
  ~ (
    slice_sample(snowfall, prop = 1, replace = TRUE) %>%
      summarize(
        mean_snow = mean(snowfall, na.rm = TRUE)
      )
  ) # Close off the tilde ~ ()
) # Close off the map_df

# Now make the histogram plots for the mean and standard deviation:
ggplot(bootstrap_samples) +
  geom_histogram(aes(x = mean_snow)) +
  ggtitle("Distribution of sample mean")
```

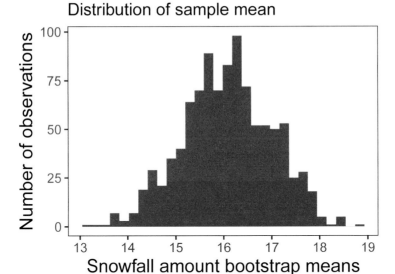

FIGURE 11.4 Histogram of 1000 bootstrap samples for the mean of the snowfall dataset.

Excellent! This is shaping up nicely. Once we have sampled as much we want, *then* investigate the distribution of the computed sample statistics (we call this the *sampling distribution*). It turns out that the statistics of the sampling distribution (such as the mean or the standard deviation) will approximate the population distribution statistics when the number of bootstrap samples is large (and 1000 is sufficiently large in this instance).

"Summarize" is the final step, where we compute summary statistics of the distribution of bootstrap means. We will do this by computing a confidence interval, which comes from the percentiles of the distribution.

Here is how we would compute these different statistics using the `quantile` command shown below:

```
# Compute the average of the distribution:
mean(bootstrap_samples$mean_snow)
```

```
## [1] 16.07825
```

```
# Compute the 2.5% and 97.5% of the distribution:
quantile(bootstrap_samples$mean_snow, probs = c(0.025, .975))
```

```
##      2.5%    97.5%
## 14.26250 17.81938
```

Notice how we used the `probs=c(0.025,.975)` command to compute the different 2.5% and 97.5% quantiles of the distribution of the sample means. Let's discuss the distribution of bootstrap means. The 2.5 percentile is approximately 14.3 inches. This means 2.5% of the distribution is at 14.3 inches or less. The 97.5 percentile is approximately 17.8 inches, so 97.5% of the distribution is 17.8 inches or less. If we take the difference between 2.5% and 97.5% that is 95%, so 95% of the distribution is contained between 14.3 and 17.8 inches. If we are using the bootstrap mean, we would report that the average precipitation is 16.1 inches with a 95% confidence interval of 14.3 to 17.8 inches. The confidence interval is to give some indication of the uncertainty in the measurements.

11.3 Summary and next steps

The idea of sampling with replacement, generating a parameter estimate, and then repeating over several iterations is at the heart of many computational parameter estimation methods, such as Markov Chain Monte Carlo methods that we will explore in Chapter 13. Bootstrapping (and other sampling with replacement techniques) makes nonlinear problems more tractable.

This chapter extended your abilities in R by showing you how to generate histograms, sample a dataset, and compute statistics. The goal here is to give you examples that you can re-use in this chapter's exercises. Enjoy!

11.4 Exercises

Exercise 11.1. Histograms are an important visualization tool in descriptive statistics. Read the following essays on histograms, and then summarize 2-3 important points of what you learned reading these articles.

- Visualizing histograms[3]
- How histograms work[4]
- How to read histograms and use them in R[5]

Exercise 11.2. Display the bootstrap histogram of 1000 bootstrap samples for the standard deviation of the snowfall dataset. From this bootstrap distribution (for the standard deviation) what is the mean and 95% confidence interval?

Exercise 11.3. (Inspired by Devore, Berk, and Carlton (2021)) Average snow

---

[3] http://tinlizzie.org/histograms/
[4] https://flowingdata.com/2017/06/07/how-histograms-work/
[5] https://flowingdata.com/2014/02/27/how-to-read-histograms-and-use-them-in-r/

cover from 1970 - 1979 in October over Eurasia (in million km^2) was reported as the following:

$$\{6.5, 12.0, 14.9, 10.0, 10.7, 7.9, 21.9, 12.5, 14.5, 9.2\}$$

a. Create a histogram for these data.
b. Compute the sample mean and median of this dataset.
c. What would you report as a representative or typical value of snow cover for October? Why?
d. The 21.9 measurement looks like an outlier. What is the sample mean excluding that measurement?

Exercise 11.4. Consider the equation $S(\theta) = (3 - 1.5^{1/\theta})^2$ for $\theta > 0$. This function is an idealized example for the cost function in Figure 11.1.

a. What is $S'(\theta)$?
b. Make a plot of $S'(\theta)$. What are the locations of the critical points?
c. Algebraically solve $S'(\theta) = 0$. Does your computed critical point match up with the graph?

Exercise 11.5. Repeat the bootstrap sample for the mean of the snowfall dataset (snowfall) where the number of bootstrap samples is 10,000. Report the median and confidence intervals for the bootstrap distribution. What do you notice as the number of bootstrap samples increases?

Exercise 11.6. Using the data in Exercise 11.3, do a bootstrap sample with $N = 1000$ to compute the a bootstrap estimate for the mean October snowfall cover in Eurasia. Compute the mean and 95% confidence interval for the bootstrap distribution.

```
       snowfall
 Min.    :11.00
 1st Qu.:13.00
 Median :15.75
 Mean    :16.10
 3rd Qu.:19.12
 Max.    :22.00
```

FIGURE 11.5 Example computing a confidence interval with the 'summary' command.

Exercise 11.7. We computed the 95% confidence interval using the quantile command. An alternative approach to summarize a distribution is with the summary command, as shown in Figure 11.5. We call this command using summary(data_frame), where data_frame is the particular data frame you want to summarize. The output reports the minimum and maximum values of a dataset. The output 1st Qu. and 3rd Qu. are the 25th and 75th percentiles.

Do 1000 bootstrap samples using the data in Exercise 11.3 and report the output from the summary command.

Exercise 11.8. The dataset precipitation lists rainfall data from a fall storm that came through the Twin Cities.

a. Make an appropriately sized histogram for the precipitation observations.
b. What is the mean precipitation across these observations?
c. Do a bootstrap estimate with $N = 100$ and $N = 1000$ and plot their respective histograms.
d. For each of your bootstrap samples ($N = 100$ and $N = 1000$) compute the mean and 95% confidence interval for the bootstrap distribution.
e. What would you report for the mean and its 95% confidence interval for this storm?

12

The Metropolis-Hastings Algorithm

Cost or likelihood functions (Chapters 9 and 10) are a powerful approach to estimate model parameters for a dataset. Bootstrap sampling (Chapter 11) is an efficient computational method to extend the reach of a dataset to estimate population level parameters. With all these elements in place we will discuss a powerful algorithm that will efficiently sample a likelihood function to estimate parameters for a model. Let's get started!

12.1 Estimating the growth of a dog

In Chapter 2 we introduced the dataset wilson, which reported data on the weight of the puppy Wilson[1] as he grew, shown again in Figure 12.1. Because we will re-use the scatter plot in Figure 12.1 several times, we define the variable wilson_data_plot so we don't have to re-copy the code over and over.[2]

```
wilson_data_plot <- ggplot(data = wilson) +
  geom_point(aes(x = days, y = weight), size = 1, color = "red") +
  labs(
    x = "D (Days since birth)",
    y = "W (Weight in pounds)"
  )

wilson_data_plot
```

---

[1] http://bscheng.com/2014/05/07/modeling-logistic-growth-data-in-r/

[2] Making a base plot and repeatedly adding to it is a really good coding practice for R.

```
# Return the likelihood from the list:
out_values
```

```
## # A tibble: 2 x 5
##      p1    p2      p3   l_hood log_lik
##   <dbl> <dbl>   <dbl>    <dbl> <lgl>
## 1    78  2.46  0.0170 6.86e-29 FALSE
## 2    65  2.46  0.0170 1.20e-27 FALSE
```

Hopefully this code seems familiar to you from Chapter 9, but of note are the following:

- We want to compare two values of `p1`, so when we defined `parameters` we included the two values of p_1 when defining `parameters`. The same values of `p2` and `p3` will apply to both. Don't believe me? Type `parameters` at the console line to see!
- Recall that when we apply `compute_likelihood` a list is returned (`likelihood` and `opt_value`). For this case we just want the `likelihood` data frame, hence the code `$likelihood` at the end of `compute_likelihood`.

So we computed $L(78)=6.8560765 \times 10^{-29}$ and $L(65)=1.2038829 \times 10^{-27}$. Since $L(65) > L(78)$ we would say p_{65} is the more likely parameter value.

Notice how we computed $L(65)$ and $L(78)$ separately and then compared the two values. Another approach is to examine the *ratio* of the likelihoods (Equation (12.3)):

$$\text{Likelihood ratio: } \frac{L(p_1^{proposed})}{L(p_1^{current})} \tag{12.3}$$

The utility of the likelihood ratio is that we can say that if the likelihood ratio is *greater* than 1, $p_1^{proposed}$ is preferred. If this ratio is *less* than 1, $p_1^{current}$ is preferred.

Applying Equation (12.3) with $p_1^{proposed} = 65$ and $p_1^{current} = 78$, we have $\frac{L(65)}{L(78)} = 18$, further confirming $p_1 = 65$ is more likely compared to the value of $p_1 = 78$.

12.2.1   Iterative improvement with likelihood ratios

We can improve on estimating p_1 for Equation (12.1) by continuing to compute likelihood ratios. However, since $p_1 = 65$ is the more likely value (currently), then we will set $p_1^{current} = 65$ for Equation (12.3).

To simplify things, let's define a function that will quickly compute Equation (12.1) for this dataset:

```r
# A function that computes the likelihood ratio for Wilson's weight
likelihood_ratio_wilson <- function(proposed, current) {

  # Define the model we are using
  my_model <- weight ~ p1 / (1 + exp(p2 - p3 * days))


  # This allows for all the possible combinations of parameters
  parameters <- tibble(
    p1 = c(current, proposed),
    p2 = 2.461935,
    p3 = 0.017032
  )

  # Compute the likelihood and return the likelihood from the list
  out_values <-
    compute_likelihood(my_model, wilson, parameters)$likelihood

  # Return the likelihood from the list:
  ll_ratio <- out_values$l_hood[[2]] / out_values$l_hood[[1]]

  return(ll_ratio)
}


# Test the function out:
likelihood_ratio_wilson(65, 78)
```

```
## [1] 17.55936
```

You should notice that the reported likelihood ratio matches up with our earlier computations! Perhaps a better guess for p_1 would be somewhere between 65 and 78. Let's try to compute the likelihood ratio for $p_1 = 70$ compared to $p_1 = 65$. Try computing `likelihood_ratio_wilson(70,65)` - you should see that it is about 7.5 million times more likely!

I think we are onto something - Figure 12.2 compares the modeled values of Wilson's weight for the different parameters:

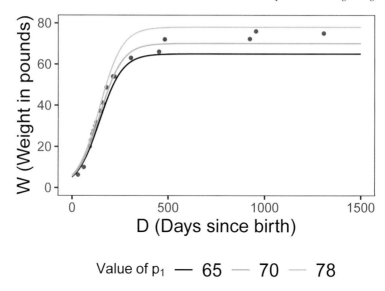

FIGURE 12.2 Comparison of our three estimates for Wilson's weight over time.

So now, let's try $p_1 = 74$ and compare the likelihoods: $\dfrac{L(74)}{L(70)} = 4.5897793 \times 10^{-9}$. This seems to be *less* likely because the ratio was significantly less than one. If we are doing a hunt for the *best* optimum value, then perhaps we would reject $p_1 = 74$ and keep moving on, perhaps selecting another value closer to 70.

While rejecting $p_1 = 74$ as less likely, a word of caution is warranted. For non-linear problems we want to be extra careful that we do not accept a parameter value that leads us to a *local* (not global) optimum. A way to avoid this is to compare the calculated likelihood ratio to a uniform random number r between 0 and 1.

At the R console type `runif(1)` - this creates one random number from the uniform distribution (remember the default range of the uniform distribution is $0 \leq p_1 \leq 1$). The `r` in `runif(1)` stands for *random*. When I tried `runif(1)` I received a value of 0.126. Since the likelihood ratio is smaller than the random number I generated, we will *reject* the value of $p_1 = 74$ and try again, keeping 70 as our value.

The process to keep the proposed value based on some decision metric is called a *decision step*.

12.3 The Metropolis-Hastings algorithm for parameter estimation

Section 12.2 outlined an iterative approach of applying likelihood ratios to estimate p_1. Let's organize all the work in a table (Table 12.1).

TABLE 12.1 Organizational table of Metropolis-Hastings algorithm to estimate p_1 from the `wilson` dataset.

Iteration	Current value of p_1	Proposed value of p_1	$\dfrac{L(p_1^{proposed})}{L(p_1^{current})}$	Value of runif(1)	Accept proposed p_1?
0	78	NA	NA	NA	NA
1	78	65	17.55936	NA	yes
2	65	70	7465075	NA	yes
3	70	74	0.09985308	0.126	no
4	70	...	...	...	...

Table 12.1 is the essence of what is called the Metropolis-Hastings algorithm[3]. The goal of this algorithm method is to determine the parameter set that optimizes the likelihood function, or makes the likelihood ratio greater than unity. Here are the key components for this algorithm:

1. A defined likelihood function.
2. A starting value for your parameter.
3. A proposed value for your parameter.
4. Comparison of the likelihood ratios for the proposed to the current value (Likelihood ratio: $\dfrac{L(p_1^{proposed})}{L(p_1^{current})}$). Parameter values that increase the likelihood will be preferred.
5. A decision to accept the proposed parameter value. If the likelihood ratio is greater than 1, then we accept this value. However if the likelihood ratio is less than 1, we generate a random number r (using runif(1)) and use this following process:

- If r is less than the likelihood ratio we **accept** (keep) the proposed parameter value.
- If r is greater than the likelihood ratio we **reject** the proposed parameter value.

---

[3]https://en.wikipedia.org/wiki/Metropolis%E2%80%93Hastings_algorithm

Exercise 12.6. An alternative model for the dog's mass is the following differential equation:

$$\frac{dW}{dt} = -k(W - p_1) \tag{12.4}$$

a. Apply separation of variables and $W(0) = 5$ and the value of p_1 from Exercise 12.4 to determine the solution for this differential equation.
b. Apply 10 iterations of the Metropolis-Hastings algorithm to estimate the value of k to three decimal places accuracy. The true value of k is between 0 and 1.
c. Compare your final estimated value of k with the data in one plot.

Exercise 12.7. Consider the linear model $y = 6.94 + bx$ for the following dataset:

x	-0.365	-0.14	-0.53	-0.035	0.272
y	6.57	6.78	6.39	6.96	7.20

Apply 10 iterations of the Metropolis-Hastings algorithm to determine b.

Exercise 12.8. For the wilson dataset, repeat three steps of the parameter estimation to determine p_1 as in this chapter, but this time use log_likelihood_ratio_wilson to estimate p_1. Which function (likelihood_ratio_wilson or log_likelihood_ratio_wilson) do you think is easier in practice?

13

Markov Chain Monte Carlo Parameter Estimation

We have explored likelihood functions, iterative methods, and the Metropolis-Hastings algorithm. In this chapter all these together introduce a sophisticated parameter estimation algorithm called Markov Chain Monte Carlo (MCMC) parameter estimation, which has a rich history (Richey 2010). MCMC methods can be highly computational; more importantly you already have the skills in place to understand *how* the MCMC method works. To do the heavy lifting we will rely on functions from the demodelr package. Let's get started!

13.1 The recipe for MCMC

The MCMC approach is a systematic exploration to determine the set of parameters that optimizes the value of the log-likelihood function, given the data. It may be helpful to think of the MCMC method as a recipe, and in order to "run" the MCMC method, you will need four key ingredients:

- *Model*: a function that we have for our dynamics (this is $\frac{d\vec{y}}{dt} = f(\vec{y}, \vec{\alpha}, t)$), or an empirical equation $\vec{y} = f(\vec{x}, \vec{\alpha})$.
- *Data*: a data frame (tibble) or a spreadsheet file (to read into R) of the data you wish to use for parameter estimation.
- *Parameter bounds*: upper and lower bounds on your parameter values. We typically assume an initial uniform distribution on the parameters.
- *Initial conditions*: (optional) needed if your model is a differential equation.
- *Run diagnostics*: specifications for the MCMC code, which may include how long you will run the code and other aspects of the MCMC algorithm.

We will work step by step through two examples of an application of the MCMC algorithm using both a differential equation and an empirical model. Example code is provided so you can also run your own estimates. The workflow that we will use is:

Define the model, parameters, and data → Determine MCMC
settings → Compute MCMC estimate → Analyze results.

Having an established workflow helps to breakdown the process step by step,
making it easier to check for any coding errors.

13.2   MCMC parameter estimation with an empirical model

The first step of our workflow is to "Define the model and parameters." Here we
return to the problem exploring of the phosphorous content in algae (denoted
by x) compared to the phosphorous content in daphnia (denoted by y), and
estimating c and θ from Equation (13.1).

$$y = c \cdot x^{1/\theta} \tag{13.1}$$

The parameters c and θ from Equation (13.1) range from $0 \leq c \leq 2$ and
$1 \leq \theta \leq 20$. To define the model we use similar code to how we defined models
in Chapter 8. Then to define the parameters we will use a `tibble`, specifying
the upper and lower bounds:

```
## Step 1: Define the model and parameters
phos_model <- daphnia ~ c * algae^(1 / theta)

# Define the parameters that you will use with their bounds
phos_param <- tibble(
  name = c("c", "theta"),
  lower_bound = c(0, 1),
  upper_bound = c(2, 20)
)
```

Notice how we defined that `tibble` called `phos_param`, which has three columns:
`name` which contains the name of the variables in our model (`c` and `theta`), the
lower (`lower_bound`) and upper (`upper_bound`) for the parameters (listed in the
same order as the parameters listed in `name`).

The data that we use is the dataset `phosphorous`, which is already located in
the `demodelr` package).

The next two steps in our workflow (Determine MCMC settings → Compute MCMC estimate) are combined together below:

```
## Step 2: Determine MCMC settings
# Define the number of iterations
phos_iter <- 1000

## Step 3: Compute MCMC estimate
phos_mcmc <- mcmc_estimate(
  model = phos_model,
  data = phosphorous,
  parameters = phos_param,
  iterations = phos_iter
)
```

The variable `phos_iter` specifies how many iterations we will run of the MCMC method. Notice that `mcmc_estimate` has several inputs, which for convenience we write on separate lines. There are four required inputs to the function `mcmc_estimate` and several predefined inputs; you will explore these further in Exercise 13.3.

The function `mcmc_estimate` may take some time (which is OK). But once it finishes a `tibble` is produced, which we call `phosphorous_mcmc` (run this code on your own):

```
glimpse(phos_mcmc)
```

Notice `phos_mcmc` contains four columns:

- `accept_flag` tells you if at that particular iteration the MCMC estimate was accepted or not. This is a categorical variable of `TRUE` or `FALSE`
- `l_hood` is the value of the likelihood for that given iteration.
- The values of the parameters follow on the next few lines. Notice that θ is written as `theta` in the resulting data frame.

The final step of our workflow is to "Analyze results." Fortunately the `demodelr` package has a function called `mcmc_analyze` to help you:

```
## Step 4: Analyze results:
mcmc_analyze(
  model = phos_model,
  data = phosphorous,
  mcmc_out = phos_mcmc
)
```

The function `mcmc_analyze` filters `phos_mcmc` whenever the variable `accept_flag` is `TRUE`. This function will compute parameter statistics (e.g. median and 95% confidence intervals) to be displayed at the console. In addition this

13.3   MCMC parameter estimation with a differential equation model

Next let's try parameter estimation with a differential equation model. Here the measured data are solutions to a differential equation, which contains unknown parameters. Once the MCMC method proposes a parameter, then the differential equation needs to be solved numerically with Euler's or a Runge-Kutta method before evaluating the likelihood function.[3]

The example that we are going to use relates to land use management, in particular a coupled system between a resource (such as a national park) and the number of visitors it receives (Sinay and Sinay 2006). The tourism model relies on two nondimensional scaled variables, R which is the amount of the resource (as a percentage) and V the percentage of visitors that could visit (also as a percentage):

$$
\begin{aligned}
\frac{dR}{dt} &= R \cdot (1 - R) - aV \\
\frac{dV}{dt} &= b \cdot V \cdot (R - V)
\end{aligned}
\tag{13.2}
$$

Equation (13.2) has two parameters a and b, which relate to how the resource is used up as visitors come (a) and how as the visitors increase, word of mouth leads to a negative effect of it being too crowded (b).

For this case we are going to use a pre-defined dataset of the number of resources and visitors to a national park as reported in Sinay and Sinay (2006) (this is the `parks` dataset in the `demodelr` package) which is plotted in Figure 13.3.

---

[3]If we knew the function that solves the differential equation, then we would have an empirical model.

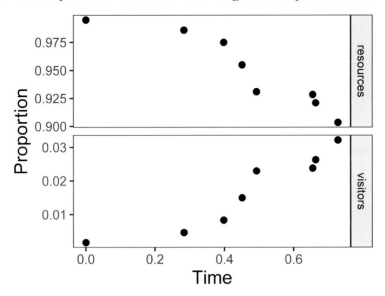

FIGURE 13.3 Scaled data on resources and visitors to a national park over time.

As the visitors V increase in Figure 13.3, the percentage of the resources R decreases. Notably though, the data on both variables may be limited. In Figure 13.3 the proportion of visitors ranges from 0.9 to 1, and the resources ranges up to 0.03. Perhaps from this limited dataset given we can estimate the parameters a and b and then forecast out as time increases. We will estimate the parameters a and b in Equation (13.2) with the data shown in Figure 13.3. We are going to assume that $0 \leq a \leq 30$ and $0 \leq b \leq 5$. We will still use the same workflow (Define the model, parameters, and data → Determine MCMC settings → Compute MCMC estimate → Analyze results) as we did in estimating parameters for an empirical model. Since this workflow was presented earlier we will combine the first three steps below:

```
## Step 1: Define the model, parameters, and data
# Define the tourism model
tourism_model <- c(
  dRdt ~ resources * (1 - resources) - a * visitors,
  dVdt ~ b * visitors * (resources - visitors)
)

# Define the parameters that you will use with their bounds
tourism_param <- tibble(
  name = c("a", "b"),
  lower_bound = c(10, 0),
  upper_bound = c(30, 5)
```

```
)

## Step 2: Determine MCMC settings
# Define the initial conditions
tourism_init <- c(resources = 0.995, visitors = 0.00167)

deltaT <- .1 # timestep length
n_steps <- 15 # must be a number greater than 1

# Define the number of iterations
tourism_iter <- 1000

## Step 3: Compute MCMC estimate
tourism_out <- mcmc_estimate(
  model = tourism_model,
  data = parks,
  parameters = tourism_param,
  mode = "de",
  initial_condition = tourism_init,
  deltaT = deltaT,
  n_steps = n_steps,
  iterations = tourism_iter
)
```

Notice how mcmc_estimate has some additional arguments. Most important is
the option mode "de", where de stands for *differential equation*. (The default
mode is emp, or *empirical* model - like the phosphorous data set.) If the de mode
is specified, then you also need to define the initial conditions (tourism_init),
Δt (deltaT), and timesteps (n_steps) in order to generate the numerical
solution.

Visualizing the data also is done with mcmc_analyze:

```
## Step 4: Analyze results
mcmc_analyze(
  model = tourism_model,
  data = parks,
  mcmc_out = tourism_out,
  mode = "de",
  initial_condition = tourism_init,
  deltaT = deltaT,
  n_steps = n_steps
)
```

Examining the parameter histograms (Figure 13.4) shows b to be well-constrained. The histogram for a seems like it could be well-constrained - but we may need to run more iterations to confirm this.

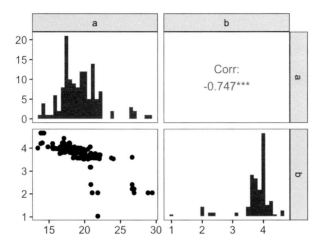

FIGURE 13.4 Pairwise parameter histogram of MCMC parameter estimation results with Equation 13.2.

The model results and confidence intervals show good agreement to the data (Figure 13.5). Additionally the model forecasts out in time confirming that as visitors increase, the resources in the national park will decrease due to overuse. In contrast to Figure 13.2, the black line in Figure 13.5 represents the median and the grey shading is the 95% confidence interval for all timesteps defined in solving the model.

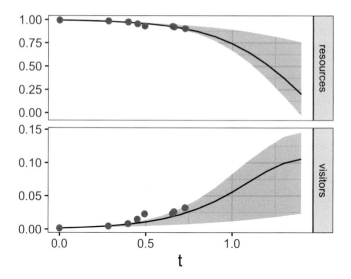

FIGURE 13.5 Ensemble output results from the MCMC parameter estimation for Equation 13.2.

13.4   Timing your code

As you can imagine the more iterations we have the better our parameter estimates will be. However, this means the full estimate with that number of iterations will take some more time. Before doing that, you first should get an estimate for the length of time it takes to run this code. Fortunately R has a stopwatch function. Let's check this out with one iteration of the phosphorous dataset:

```r
# This "starts" the stopwatch
start_time <- Sys.time()

# Compute a single mcmc estimate
phosphorous1_mcmc <- mcmc_estimate(
  model = phos_model,
  data = phosphorous,
  parameters = phos_param,
  iterations = 1
)

# End the stopwatch
end_time <- Sys.time()
```

```
# Determine the difference between the start and end times
end_time - start_time
```

```
## Time difference of 0.231061 secs
```

Timing the code for one iteration gives you a ballpark estimate for a full MCMC parameter estimate. If we were to run N MCMC iterations, a good benchmark would be to multiply the time difference (`end_time - start_time`) by N. Performance time varies by computer and the other programs / apps that are running at the same time. However, this gives you an estimate of what to expect.[4]

13.5 Further extensions to MCMC

For the examples in this chapter we limited the number of iterations to a smaller number to make the results computationally feasible. However we can extend the MCMC approach in two notable ways:

- One approach is to separate the data into two different sets - one for optimization and one for validation. In this approach the "optimization data" consists of a certain percentage of the original dataset, leaving the remaining to validate the forward forecasts. This is a type of cross-validation approach, and is generally preferred because you are demonstrating the strength of your model ability against non-optimized data.

- We also run multiple "chains" of optimization, starting from a different value in parameter space. What we do then after running each of these chains is to select the one with the best log-likelihood value, and run *another* MCMC iteration starting at that value. The idea is with a different chain we have sampled the parameter space and are hopefully starting near an optimum value.

As you can see, the MCMC algorithm is an extremely powerful technique for parameter estimation. While MCMC may take additional time and programming skill to analyze - it is definitely worth it!

---

[4]I am a big fan of "set it and forget it" - meaning I set up the code before I go to sleep and it is ready in the morning!

13.6   Exercises

Exercise 13.1. Re-run both of the MCMC examples in this chapter, but increase the number of iterations to $10,000$. Analyze your results from both cases. How does increasing the number of iterations affect the posterior parameter estimates and their confidence intervals? Does the log-likelihood value change?

Exercise 13.2. Time the MCMC parameter estimate for the phosphorous dataset for 1 iteration. Then time the MCMC parameter estimate for 10, 100, and 1000 iterations, recording the times for each one. Make a scatterplot with the number of iterations on the horizontal axis and time on the vertical axis. How would you characterize the relationship between the number of iterations and the time it takes to run the code? How long would it take to compute an MCMC estimate with $10,000$ iterations?

Exercise 13.3. The function mcmc_estimate has several other input variables that are set to default values. What are they and how would you explain their use? (*Hint:* to see the documentation associated with this function type ?mcmc_estimate at the R console.)

Exercise 13.4. For the parks data (Equation (13.2)) studied in this chapter, compare the 1:1 and the posterior parameter plots (Figure 13.4). Write a summary of each panel of the plot. Apply your understanding of equifinality and other observations to determine by how much you have estimated the parameters a and b from the data.

Exercise 13.5. Run an MCMC parameter estimation on the dataset yeast from Gause (1932), where the equation for the volume of yeast V over time is given by the following equation for a yeast growing in isolation:

$$V = \frac{K}{1 + e^{a - bt}}, \tag{13.3}$$

where K is the carrying capacity, a and b respective rate constants.

a. Show that when $V(0) = 0.45$, $a = \ln(K/0.45 - 1)$.
b. Rewrite the $V(t)$ equation without a.
c. With the yeast data, perform an MCMC estimate for this equation. (*Reminder:* $\ln(5)$) is implemented as log(5) in R.)

Use the following settings for your MCMC parameter estimation:

- K : 1 to 20
- b: 0 to 1

- 1000 iterations

When setting up the MCMC method, be sure to name the variables in your model to match the yeast data frame.

d. Report all outputs from the MCMC estimation (this includes parameter estimates, confidence intervals, log-likelihood values, and any graphs). Compare your results to the results from Exercise 13.6.

Exercise 13.6. Another model for this growth of yeast is the function $V = K + Ae^{-bt}$.

a. Show that when $V(0) = 0.45$, $A = K - 0.45$.
b. Rewrite the initial equation without A.
c. With the yeast data, apply an MCMC estimate for this equation.

Use the following settings for your MCMC parameter estimation:

- K : 1 to 20
- b: 0 to 1
- 1000 iterations

When setting up the MCMC method, be sure to name the variables in your model to match the yeast data frame.

d. Report all outputs from the MCMC estimation (this includes parameter estimates, confidence intervals, log-likelihood values, and any graphs). Compare your results to the results from Exercise 13.5.

Exercise 13.7. Run an MCMC parameter estimation on the dataset wilson according to the following differential equation:

$$\frac{dP}{dt} = b(N - P), \tag{13.4}$$

where P represents the mass of the dog. Use the following settings for your MCMC parameter estimation:

- N: 60 to 90
- b: 0 to .01
- $P(0) = 5$
- $\Delta t = 1$ day
- Number of timesteps: 1500
- Number of iterations: 1000

When setting up the MCMC method, be sure to name the variables in your model to match the wilson data frame. Be sure to report all outputs from the MCMC estimation (this includes parameter estimates, confidence intervals, log-likelihood values, and any graphs).

14

Information Criteria

In Exercises 13.5 and 13.6 of Chapter 13 we introduced two different empirical models for fitting the growth of yeast V over time t. One model is a logistic model ($V = \dfrac{K}{1 + e^{a-bt}}$), whereas the second model is a saturating function ($V = K + Ae^{-bt}$). A plot comparing MCMC parameter estimates for the two models is shown in Figure 14.1.

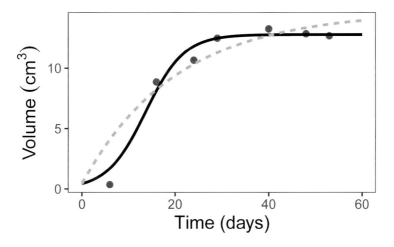

FIGURE 14.1 Comparison of models for the growth of yeast in culture. Dots represent measured values from Gause (1932).

Figure 14.1 raises an interesting question. Sometimes we have multiple, convergent models to describe a context or situation. While having these different options is good, we also like to know which is the *best* model. How would you decide that?

This chapter focuses on objective criteria to assess what is called the *best approximating model* (Burnham and Anderson 2002). We will explore what are called *information criteria*, which is developed from statistical theory. Let's get started!

14.1   Model assessment guidelines

The first step is to develop some guidelines and metrics for model evaluation. Here would be the start of a list of things to consider, represented as questions:

- The model complexity - how many equations do we have?
- The number of parameters - a few or many?
- Do the model outputs match the data?
- How will model prediction compare to any newly collected measurements?
- Are the trends accurately represented (especially for timeseries data)?
- Is the selected model easy to use, simulate, and forecast?

I may have hinted at some of these guidelines in earlier chapters. These questions are related to one another - and answering these questions (or ranking criteria for them) is at the heart of the topic of *model selection*.

Perhaps you may be asking, why bother? Aren't more models better? Let's talk about a specific example, for which we return to the dataset `global_temperature` in the `demodelr` library. Recall this dataset represents the average global temperature anomaly relative to 1951-1980. When we did linear regression with this dataset in Chapter 8 the quadratic and cubic models were approximately the same (Figure 14.2):

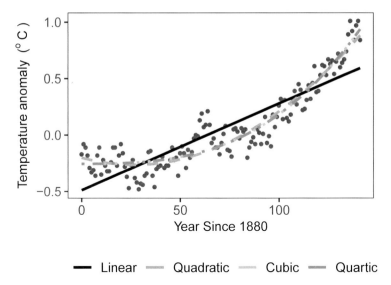

FIGURE 14.2 Comparison of global temperature anomaly dataset with various polynomial fitted models.

The variation in the different model fits for Figure 14.2 shows how different, but similar, the model results can be depending on the choice of regression function. Table 14.1 displays summary results for the log-likelihood and the root mean square error (RMSE).[1] In some cases, the log-likelihood decreases (indicating a more likely model), supported by the decrease in the RMSE indicating the fitted model more closely matches the observations. However the decrease in the log-likelihood and the RMSE is changing less as the complexity of the model (i.e. a higher degree polynomial) increases.

TABLE 14.1 Comparison of model fits for global temperature anomaly dataset shown in Figure 14.2.

Model	Log-likelihood[2]	RMSE
Linear	45.04	0.176
Quadratic	103.305	0.117
Cubic	105.788	0.115
Quartic	112.5	0.11

Further model evaluation can be examined by the following:

---

[1] The root mean square error is the computed as $\sqrt{\dfrac{\sum (y_i - f(x_i))^2}{N}}$.

[2] Remember, log-likelihoods can be positive or negative; see Chapter 9.

- Compare the measured values of $\vec{y}$ to the modeled values of $\vec{y}$ in a 1:1 plot. Does g do a better job predicting $\vec{y}$ than f?
- Related to that, compare the likelihood function values of f and g. We would favor the model that has the lower log-likelihood.
- Compare the number of parameters in each model f and g. We would favor the model that has the fewest number of parameters.

Given the above question, we can state the model selection problem as the following:

When we have two $f(\vec{x}, \vec{\alpha})$ and $g(\vec{x}, \vec{\beta})$ for the data $\vec{y}$, how would we determine which one (f or g or perhaps another alternative model) is the best approximating model?

14.2   Information criteria for assessing competing models

Information criteria evaluate the tradeoff between model complexity (i.e. the number of parameters used) and with the log-likelihood (a measure of how well the model fits the data). There are several types of information criteria, but we are going to focus on two:

- The **Akaike Information Criterion** (*AIC*, Akaike (1974)) is the most commonly used information criteria:

$$AIC = -2LL_{max} + 2P \qquad (14.1)$$

- An alternative to the *AIC* is the **Bayesian Information Criterion** (*BIC*, Schwartz (1978))

$$BIC = -2LL_{max} + P\ln(N) \qquad (14.2)$$

In Equations (14.1) and (14.2), N is the number of data points, P is the number of estimated parameters, and LL_{max} is the log-likelihood for the parameter set that maximized the likelihood function. Equations (14.1) and (14.2) show the dependence on the log-likelihood function and the number of parameters. For both the *AIC* and *BIC* a lower value of the information criteria indicates greater support for the model from the data.

Notice how easy the *AIC* and *BIC* are to compute in Equations (14.1) and (14.2) (assuming you have the information at hand). When an empirical model fit is computed (i.e. using the command lm), R computes these easily with the functions AIC or BIC. To apply them you need to first do the model fit (with the function lm. Try this out by running the following code on your own:[3]:

```
regression_formula <- temperature_anomaly ~ 1 + year_since_1880
fit <- lm(regression_formula, data = global_temperature)
AIC(fit)
BIC(fit)
```

Table 14.2 compares *AIC* and *BIC* for the models fitted using the global temperature anomaly dataset:

TABLE 14.2 Comparison of the *AIC* and *BIC* for global temperature anomaly dataset shown in Figure 14.2.

Model	*AIC*	*BIC*
Linear	-84.08	-75.213
Quadratic	-198.609	-186.786
Cubic	-201.576	-186.797
Quartic	-213	-195.265

Table 14.2 shows that the cubic model is the better approximating model for both the *AIC* and the *BIC*.

14.3   A few cautionary notes

- Information criteria are relative measures. In a study it may be more helpful to report the change in the information criteria, or even a ratio (see Burnham and Anderson (2002) for a detailed analysis).
- Information criteria are not cross-comparable across studies. If you are pulling in a model from another study, it is helpful to re-calculate the information criteria.
- An advantage to the *BIC* is that it measures tradeoffs between favoring a model that has the fewer number of data needed to estimate parameters. Other information criteria examine the distribution of the likelihood function and parameters.

---

[3]You can compute the log-likelihood with the function logLik(fit), where fit is the result of your linear model fits.

The upshot: Information criteria are *one* piece of evidence to help you to evaluate the best approximating model. You should do additional investigation (parameter evaluation, model-data fits, forecast values) in order to help determine the best model.

14.4   Exercises

Exercise 14.1. You are investigating different models for the growth of a yeast species in a population where V is the rate of reaction and s is the added substrate:

$$\text{Model 1: } V = \frac{V_{max}s}{s + K_m}$$
$$\text{Model 2: } V = \frac{K}{1 + e^{-a-bs}}$$
$$\text{Model 3: } V = K + Ae^{-bs}$$

With a dataset of 7 observations you found that the log-likelihood for Model 1 is 26.426, for Model 2 the log-likelihood is is 15.587, and for Model 3 the the log-likelihood is 21.537. Apply the AIC and the BIC to evaluate which model is the best approximating model. Be sure to identify the number of estimated parameters for each model.

Exercise 14.2. An equation that relates a consumer's nutrient content (denoted as y) to the nutrient content of food (denoted as x) is given by: $y = cx^{1/\theta}$, where $\theta \geq 1$ and c are parameters. We can apply linear regression to the dataset $(x, \ln(y))$, so the intercept of the linear regression equals $\ln(c)$ and the slope equals $1/\theta$.

a. Show that you can write this equation as a linear equation by applying a logarithm to both sides and simplifying.
b. With the dataset `phosphorous`, take the logarithm of the `daphnia` variable and then determine a linear regression fit for your new linear equation. What are the reported values of the slope and intercept from the linear regression, and by association, c and θ?
c. Apply the function `logLik` to report the log-likelihood of the fit.
d. What are the reported values of the AIC and the BIC?
e. An alternative linear model is the equation $y = a + b\sqrt{x}$. Use the R command `sqrt_fit <- lm(daphnia~I(sqrt(algae)),data = phosphorous)` to first obtain a fit for this model. Then compute the log-likelihood and the AIC and the BIC. Of the two models (the log-transformed model and the square root model), which one is the better approximating model?

Exercise 14.3. (Inspired by Burnham and Anderson (2002)) You are tasked with the job of investigating the effect of a pesticide on water quality, in terms of its effects on the health of the plants and fish in the ecosystem. Different models can be created that investigate the effect of the pesticide. Different types of reaction schemes for this system are shown in Figure 3.7, where F represents the amount of pesticide in the fish, W the amount of pesticide in the water, and S the amount of pesticide in the soil. The prime (e.g. F', W', and S') represent other bound forms of the respective state. In all seven different models can be derived.

These models were applied to a dataset with 36 measurements of the water, fish, and plants. The table for the log-likelihood for each model is shown below:

Model	1a	2a	2b	3a	3b	4a	4b
Log-likelihood	-90.105	-71.986	-56.869	-31.598	-31.563	-8.770	-14.238

a. Use Figure 3.7 to identify the number of parameters for each model.
b. Apply the AIC and the BIC to the data in the above table to determine which is the best approximating model.

Exercise 14.4. Use the information shown in Table 14.1 to compute (by hand) the AIC and the BIC for each of the models for the global_temperature dataset (there are 142 observations). Do your results conform to what is presented in Table 14.2? How far off are your results? What would be a plausible explanation for the difference?

Part III

Stability Analysis for Differential Equations

15

Systems of Linear Differential Equations

Here we delve into a deeper understanding of differential equations by examining long term stability of equilibrium solutions. As a first step, Chapter 15 focuses on *linear* systems of differential equations, such as Equation (15.1):

$$\frac{dx}{dt} = 2x$$
$$\frac{dy}{dt} = x + y \tag{15.1}$$

Equation (15.1) is a linear system of differential equations because it does contain terms such as y^2 or $\sin(x)$ on the right hand side of the equation. This chapter focuses on visualizing the phase plane for linear systems and determining the equilibrium solutions. Let's get started!

15.1 Linear systems of differential equations and matrix notation

Another way to express Equation (15.1) is with matrix notation:

$$\begin{pmatrix} \frac{dx}{dt} \\ \frac{dy}{dt} \end{pmatrix} = \begin{pmatrix} 2x \\ x + y \end{pmatrix}$$
$$= \begin{pmatrix} 2 & 0 \\ 1 & 1 \end{pmatrix} \begin{pmatrix} x \\ y \end{pmatrix} \tag{15.2}$$

(Note: we can also use the prime notation (x' or y') to signify $\frac{dx}{dt}$ or $\frac{dy}{dt}$.) We can also represent Equation (15.2) in a compact vector notation: $\frac{d\vec{x}}{dt} = A\vec{x}$, where for this example $\vec{A} = \begin{pmatrix} 2 & 0 \\ 1 & 1 \end{pmatrix}$.

Now let's generalize. A system of linear equations:

$$\frac{dx}{dt} = ax + by$$
$$\frac{dy}{dt} = cx + dy$$

(15.3)

can be expressed in the following way:

$$\begin{pmatrix} \frac{dx}{dt} \\ \frac{dy}{dt} \end{pmatrix} = \begin{pmatrix} ax + by \\ cx + dy \end{pmatrix} = \begin{pmatrix} a & b \\ c & d \end{pmatrix} \begin{pmatrix} x \\ y \end{pmatrix}$$

(15.4)

Equation (15.1) is an example of a *coupled* system of equations, mainly due to the expression $\frac{dy}{dt} = x + y$. An example of an *uncoupled* system of equations would be $\frac{dx}{dt} = 3x$ and $\frac{dy}{dt} = -2y$, which could be solved with separation of variables.

15.2   Equilibrium solutions

Now let's discuss equilibrium solutions for Equation (15.1). Recall equilibrium solutions are places where both $\frac{dx}{dt} = 0$ and $\frac{dy}{dt} = 0$. Since $\frac{dx}{dt} = 2x$, an equilibrium solution would be $x = 0$. Substituting $x = 0$ into $\frac{dy}{dt} = x + y$ also shows $y = 0$ is the corresponding equilibrium solution for y.

For a general linear system of differential equations, it might be helpful to imagine what we should *expect* for an equilibrium solution. Think back to calculus - what types of functions have a derivative that equals zero? (Hopefully constant functions comes to mind!) The equilibrium solution is then $x = 0$ and $y = 0$.

Here is an amazing fact: it turns out **any linear system of differential equations has the origin as its only equilibrium solution.** One way to verify this fact is to examine the theory behind solutions for linear systems of equations in linear algebra. You might be wondering why there is all the fuss with equilibrium solutions - especially the origin ($x = 0$ and $y = 0$)[1]. So while equilibrium solutions are not a terribly interesting question at the moment, the *stability* of solutions is. In order to understand what I mean by stability, let's re-examine how to generate phase planes from Chapter 6.

---

[1] Another name for the origin equilibrium solution is the *trivial equilibrium.* Can you see why it is trivial?

15.3 The phase plane

The phase plane is helpful here to understand the stability of an equilibrium solution. Remember that the phase plane shows the motion of solutions, visualized as a vector. For the system we examined earlier let's take a look at the phase plane. Here is some R code from the demodelr package to help you visualize the phase plane for Equation (15.1), shown in Figure 15.1.[2]

```r
# For a two variable system of differential equations
# we need to define dx/dt and dy/dt separately:


linear_eq <- c(
  dxdt ~ 2 * x,
  dydt ~ x + y
)

# Now we plot the solution.
phaseplane(
  system_eq = linear_eq,
  x_var = "x",
  y_var = "y"
)
```

---

[2]It is okay to refer back to Chapter 6 for a refresher on how the phaseplane command works.

16

Systems of Nonlinear Differential Equations

16.1 Introducing nonlinear systems of differential equations

In Chapter 15 we discussed systems of linear equations. For this chapter we focus on *non*-linear systems of equations. We previously discussed coupled (nonlinear) systems of equations in Chapter 6, but we will dig in a little deeper here.

Consider the following nonlinear system of equations with the associated phase plane in Figure 16.1:

$$\frac{dx}{dt} = y - 1$$
$$\frac{dy}{dt} = x^2 - 1$$

(16.1)

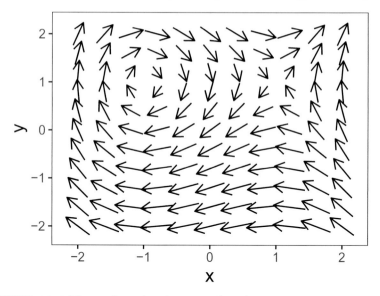

FIGURE 16.1 Phase plane for Equation (16.1).

Wow! The phase plane in Figure 16.1 looks really interesting. Let's dig into this deeper to understand the phase plane better.

16.2   Zooming in on the phase plane

One way to investigate the phase plane is to zoom in on interesting chapters for Figure 16.1. In the upper left corner there is some swirling action, so let's zoom in somewhat (remember you can adjust the window size in `phaseplane` with the option `x_window` and `y_window`):

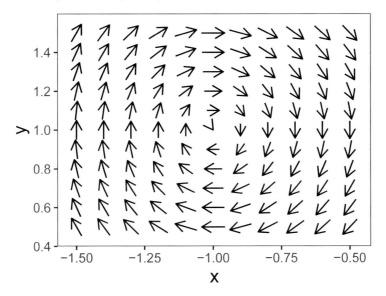

FIGURE 16.2 Zoomed in phase plane for Equation (16.1).

Something interesting seems to be happening at the point $(x, y) = (-1, 1)$ in Figure 16.2. Let's take a look at what happens if we evaluate our differential equation at $(x, y) = (-1, 1)$:

$$\begin{aligned}
\frac{dx}{dt} &= 1 - 1 = 0 \\
\frac{dy}{dt} &= (-1)^2 - 1 = 0
\end{aligned} \tag{16.2}$$

Aha! So the point $(-1, 1)$ is an equilibrium solution. In later chapters we will discuss *why* we are observing the behavior with the swirling arrows. For now, the key point from Figure 16.2 is to recognize that *nonlinear systems* can have nonzero equilibrium solutions.

Next, there seems to be a second interesting point in the upper right corner of Figure 16.1. Let's zoom in near the point $(x, y) = (1, 1)$:

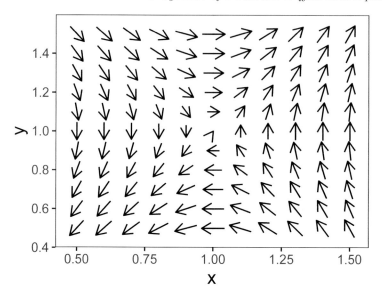

FIGURE 16.3 Another zoomed in phase plane for Equation (16.1).

It seems like there is a *second* equilibrium solution at the point $(1,1)$! Let's confirm this:

$$\frac{dx}{dt} = 1 - 1 = 0$$
$$\frac{dy}{dt} = (1)^2 - 1 = 0 \qquad (16.3)$$

By zooming in on the phase plane we learned something important about nonlinear systems and how they might differ compared to linear systems. In Chapter 15 we learned that the origin is the only equilibrium solution for a linear system of differential equations. On the other hand, nonlinear systems of equations may have *multiple* equilibrium solutions.

16.3   Determining equilibrium solutions with nullclines

To determine an equilibrium solution for a system of differential equations we first need to find the intersection of different nullclines. We do this by setting each of the rate equations ($\frac{dx}{dt}$ or $\frac{dy}{dt}$) equal to zero. Equation (16.1) has two nullclines:

$$\frac{dx}{dt} = 0 \rightarrow y - 1 = 0$$

$$\frac{dy}{dt} = 0 \rightarrow x^2 - 1 = 0 \tag{16.4}$$

So, solving for both nullclines in Equation (16.4) we have that $y = 1$ or $x = \pm 1$. You can visually see the phase plane with the nullclines in Figure 16.4, where we will add the nullclines and equilibrium solutions into the plot.

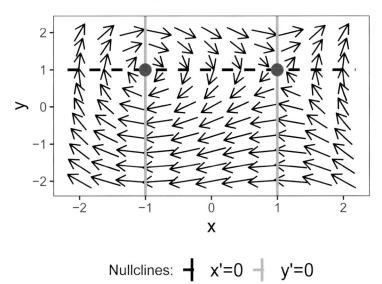

Nullclines: $\vdash$ x'=0 $\dashv$ y'=0

FIGURE 16.4 Phase plane for Equation (16.1), with nullclines and equilibrium solutions shown.

In Figure 16.4 we can see equilibrium solutions occur where a nullcline for $x' = 0$ intersects with a nullcline where $y' = 0$.

16.4 Stability of an equilibrium solution

The idea of stability of an equilibrium solution for a nonlinear system is intuitively similar to that of a linear system: the equilibrium is stable when all the phase plane arrows point towards the equilibrium solution. For Equation (16.5), the equilibrium solution at $(x, y) = (1, 1)$ is *unstable* because in Figure 16.3 some of the arrows point towards the equilibrium solution, whereas others point away from it. For Figure 16.2 it is a little harder to tell stability of the equilibrium solution at $(x, y) = (-1, 1)$. At this point we won't discuss more specifics of determining stable versus unstable equilibrium solutions. If the phase plane suggests that the equilibrium solution is stable or unstable, then you have established some good intuition that can be confirmed with additional analyses.

16.5   Graphing nullclines in a phase plane

Let's look at another example, but this time we will focus on generating graphs for the nullclines.

$$\frac{dx}{dt} = x - 0.5yx$$
$$\frac{dy}{dt} = yx - y^2 \tag{16.5}$$

Figure 16.5 shows the phase plane for this example. Can you guess where an equilibrium solution would be?

```
# Define the range we wish to use to evaluate this vector field
system_eq <- c(
  dx ~ x - 0.5 * y * x,
  dy ~ y * x - y^2
)


p1 <- phaseplane(system_eq, "x", "y",
                 x_window = c(0, 4),
                 y_window = c(0, 4)
                 )


p1
```

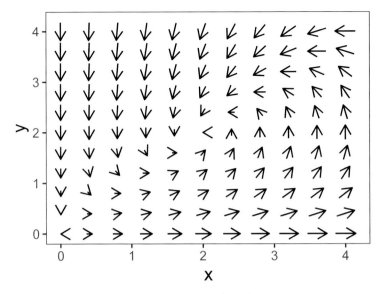

FIGURE 16.5 Phase plane for Equation (16.5).

Notice how the code used to generate Figure 16.5 stores the phase plane in the variable p1 and the displays it. This will make things easier when we plot the nullclines. Speaking of nullclines, let's find them:

$$\frac{dx}{dt} = 0 \rightarrow x - 0.5yx = 0$$
$$\frac{dy}{dt} = 0 \rightarrow yx - y^2 = 0 \tag{16.6}$$

The algebra is becoming a little more involved. Factoring $x - 0.5yx = 0$ we have $x \cdot (1 - 0.5y) = 0$, so either $x = 0$ or $y = 2$. Factoring the second equation we have $y \cdot (x - y) = 0$, so either $y = 0$ or $x = y$. Notice how this second nullcline is a function of x and y. The following code plots the phase plane (p1) along with the nullclines (try this code out on your own):

```
# Define the nullclines for dx/dt = 0 (red):

# x = 0
nullcline_x1 <- tibble(x = 0,
                       y=seq(0,4,length.out=100)
)

# y = 0.5
nullcline_x2 <- tibble(x = seq(0,4,length.out=100),
                       y=2
```

```
)

# Define the nullclines for dy/dt = 0 (blue):
# y = 0
nullcline_y1 <- tibble(x = seq(0,4,length.out=100),
                       y=0
)

# y = x
nullcline_y2 <- tibble(x = seq(0,4,length.out=100),
                       y=x
)

# Add the nullclines onto the phase plane
p1 +
   geom_line(data = nullcline_x1,aes(x=x,y=y),color='red') +
   geom_line(data = nullcline_x2,aes(x=x,y=y),color='red') +
   geom_line(data = nullcline_y1,aes(x=x,y=y),color='blue') +
   geom_line(data = nullcline_y2,aes(x=x,y=y),color='blue')
```

For each nullcline we define a data frame (`tibble`) that encodes the relevant information so we can plot it. In order to accomplish this we defined a sequence of values ranging from the plot window of 0 to 4 for the *other* variable. For nullclines where y was a function of x we defined a sequence of values for x and defined y accordingly.

For Equation (16.5) the equilibrium solutions are $(x, y) = (0, 0)$, $(x, y) = (2, 2)$. You may be tempted to think that $(0, 2)$ is also an equilibrium solution - however - $x = 0$ and $y = 2$ are equations for the x nullcline. It is easy to forget, but equilibrium solutions are determined from the intersection of *distinct* nullclines.

Now that we have seen how nonlinear systems are different from linear systems, Chapter 17 will introduce tools for analysis for the stability of equilibrium solutions.

16.6 Exercises

Exercise 16.1. Equation (16.5) equilibrium solutions are $(x, y) = (0, 0)$, $(x, y) = (2, 2)$. Zoom in on the phase plane at each of those points to determine the stability of the equilibrium solutions. (Set the window between $-0.5 \leq x \leq 0.5$ and $-0.5 \leq x \leq 0.5$ for the $(x, y) = (0, 0)$ equilibrium solution.)

Exercise 16.2. Consider the following nonlinear system of equations, which is a modification of Equation (16.1):

$$\frac{dx}{dt} = y - x$$
$$\frac{dy}{dt} = x^2 - 1 \qquad\qquad (16.7)$$

a. What are the equations for the nullclines for this differential equation?
b. What are the equilibrium solutions for this differential equation?
c. Generate a phase plane that includes all equilibrium solutions (use the window $-2 \le x \le 2$ and $-2 \le y \le 2$)
d. Based on the phase plane, evaluate the stability of the equilibrium solution.

Exercise 16.3. Consider the following nonlinear system of equations:

$$\frac{dx}{dt} = x - .5xy$$
$$\frac{dy}{dt} = .5yx - y \qquad\qquad (16.8)$$

a. What are the equations for the nullclines for this differential equation?
b. What are the equilibrium solutions for this differential equation?
c. Generate a phase plane that includes all equilibrium solutions.
d. Based on the phase plane, evaluate the stability of the equilibrium solution.

Exercise 16.4. (Inspired by Logan and Wolesensky (2009)) A population of fish F has natural predators P. A model that describes this interaction is the following:

$$\frac{dF}{dt} = F - .3FP$$
$$\frac{dP}{dt} = .5FP - P \qquad\qquad (16.9)$$

a. What are the equations for the nullclines for this differential equation?
b. What are the equilibrium solutions for this differential equation?
c. Generate a phase plane that includes all the equilibrium solutions.
d. Based on the phase plane, evaluate the stability of the equilibrium solution.

Exercise 16.5. Consider the following system:

$$\frac{dx}{dt} = y^2$$
$$\frac{dy}{dt} = -x \qquad\qquad (16.10)$$

a. What are the nullclines for this system of equations?

b. What is the equilibrium solution for this system of equations?
c. Generate a phase plane that includes the equilibrium solution. Set the viewing window to be $-0.5 \leq x \leq 0.5$ and $-0.5 \leq y \leq 0.5$.
d. Based on the phase plane, evaluate the stability of the equilibrium solution.

Exercise 16.6. The *Van der Pol Equation* is a second-order differential equation used to study radio circuits: $x'' + \mu \cdot (x^2 - 1)x' + x = 0$, where μ is a parameter.

a. Let $x' = y$ (note: $x' = \dfrac{dx}{dt}$). Show that with this change of variables the Van der Pol equation can be written as a system:

$$\frac{dx}{dt} = y$$
$$\frac{dy}{dt} = -x - \mu \cdot (x^2 - 1)y \qquad (16.11)$$

b. By determining the nullclines, verify that the only equilibrium solution is $(x, y) = (0, 0)$.
c. Make a phase plane for different values of μ ranging from -3, -1, 0, 1, 3. Set your x and y windows to range between -1 to 1.
d. Based on the phase planes that you generate, evaluate the stability of the equilibrium solution as μ changes.

Exercise 16.7. (Inspired by Strogatz (2015)) Consider the following nonlinear system:

$$\frac{dx}{dt} = y - x$$
$$\frac{dy}{dt} = -y + \frac{5x^2}{4 + x^2} \qquad (16.12)$$

a. What are the equations for the nullclines?
b. Using desmos (or some other graphing utility), graph the two nullclines simultaneously. What are the intersection points?
c. Generate a phase plane for this system that contains all the equilibrium solutions.
d. Let's say instead that $\dfrac{dx}{dt} = bx - y$, where b is a parameter such that $0 \leq b \leq 2$. Using desmos (or some other graphing utility), how many equilibrium solutions do you have as b changes?

Exercise 16.8. (Inspired by Logan and Wolesensky (2009)) Let C be the amount of carbon in a forest ecosystem, with P as the rate of increase in carbon due to photosynthesis. Herbivores H consume carbon on the following predator-prey model:

$$\frac{dC}{dt} = P - bHC$$

$$\frac{dH}{dt} = e \cdot bHC - dC$$

(16.13)

In the above equation, b, e, and d are all parameters greater than zero.

a. What are the equations for the nullclines?
b. Set $e = b = d = 1$. Plot the equations of the nullclines. How many equilibrium solutions does this system have?
c. Determine the equilibrium solutions for this system of equations, expressed in terms of the parameters b, e, and d.

17

Local Linearization and the Jacobian

Chapters 15 and 16 focused on systems of differential equations and using phase planes to determine a preliminary classification of any equilibrium solution. In this chapter we study local linearization and the associated Jacobian matrix. These tools are used to analyze stability of equilibrium solutions for a nonlinear system, thereby building a bridge between nonlinear and linear systems of equations. Let's get started!

17.1 Competing populations

Let's take a look at a familiar example from Chapter 9, specifically the experiments of growing different species of yeast together (Gause 1932). Equation (17.1) (adapted from the one presented in Gause (1932)) represents the volume of two species of yeast (which we will call Y and N) growing in the same solution:

$$\frac{dY}{dt} = .2Y \left(\frac{13 - Y - 2N}{13} \right)$$
$$\frac{dN}{dt} = .06N \left(\frac{6 - N - 0.4Y}{6} \right)$$

(17.1)

While Equation (17.1) is a tricky model to consider, the terms $2N$ and $0.4Y$ represent the effect that Y and N have on each other since they are competing for the same resource. If these terms weren't present, both Y and N would follow a logistic growth (verify this on your own).

Equation (17.1) has 4 equilibrium solutions: $(Y, N) = (0, 0)$, $(Y, N) = (13, 0)$, $(Y, N) = (0, 6)$, and $(Y, N) = (5, 4)$ (Exercise 17.1). Figure 17.1 shows the phase plane for this system along with the equilibrium solutions.[1]

---

[1] I encourage you to use the phaseplane command to re-create Figure 17.1. If you do, be sure to use the options x_window and y_window to set things correctly.

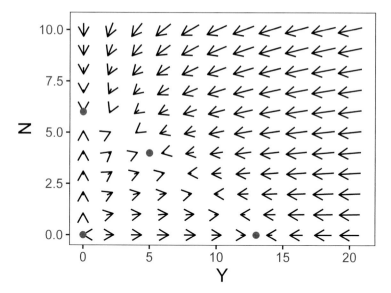

FIGURE 17.1 Phase plane for Equation (17.1), with equilibrium solutions shown as red points.

Let's take a closer look at the phase plane near the equilibrium solution $(Y, N) = (5, 4)$ in Figure 17.2:

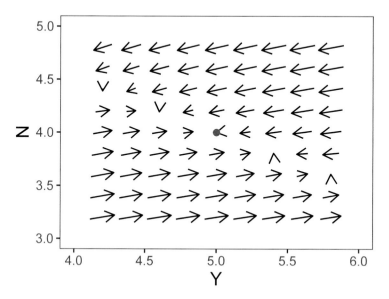

FIGURE 17.2 A zoomed in view of Equation (17.1) near the $(Y, N) = (5, 4)$ equilibrium solution.

The phase plane in Figure 17.2 *looks* like this equilibrium solution is stable (the arrows seem to suggest a "swirling" into this equilibrium solution. However, another way to verify this is by applying a **locally linear approximation**. Better stated, because Equation (17.1) is a system of differential equations, we will construct a **tangent plane approximation** around $Y = 5$, $N = 4$. Let's review how to do that next.

17.2 Tangent plane approximations

For a multivariable function $f(x, y)$, the tangent plane approximation at the point $x = a$, $y = b$ is given by Equation (17.2):

$$L(x, y) = f(a, b) + f_x(a, b) \cdot (x - a) + f_y(a, b) \cdot (y - b), \tag{17.2}$$

where f_x is the partial derivative of $f(x, y)$ with respect to x and f_y is the partial derivative of $f(x, y)$ with respect to y.

We will apply Equation (17.2) to Equation (17.1) at the equilibrium solution at $(Y, N) = (5, 4)$. Since we have *two* equations, we need to compute two tangent plane approximations (one for each equation). The right hand sides for Equation (17.1) look complicated, but we can expand them to identify $f(Y, N) = .2Y - .03YN - .015Y^2 N$ and $g(Y, N) = .06N - .01N^2 - .004YN$.

First consider $f(Y, N)$. Let's compute the partial derivatives for $f(Y, N)$ at the equilibrium solution:

$$\begin{aligned} f_Y &= .2 - .03N - .03YN \to f_Y(5, 4) = -.52 \\ f_N &= .03Y - .015Y^2 \to f_N(5, 4) = -.225 \end{aligned} \tag{17.3}$$

We also know that $f(5, 4) = 0$. As a result, the tangent plane approximation for $f(Y, N)$ is given by Equation (17.4):

$$f(Y, N) \approx -.52 \cdot (Y - 5) - .225 \cdot (N - 4) \tag{17.4}$$

Likewise if we consider $g(Y, N) = .06N - .01N^2 - .004YN$, we have:

$$\begin{aligned} g_Y &= -.004N \to g_Y(5, 4) = -.016 \\ g_N &= .06 - .02N - .004Y \to g_N(5, 4) = -.04 \end{aligned} \tag{17.5}$$

We also know that $g(5, 4) = 0$. As a result, the tangent plane approximation for $g(Y, N)$ is given by Equation (17.6):

17.4   Exercises

Exercise 17.1. Using algebra, show that the 4 equilibrium solutions to Equation (17.1) are $(Y, N) = (0, 0)$, $(Y, N) = (13, 0)$, $(Y, N) = (0, 6)$, and $(Y, N) = (5, 4)$ *Hint:* Perhaps first determine the nullclines for each solution.

Exercise 17.2. Construct the Jacobian matrices for the equilibrium solutions $(Y, N) = (13, 0)$ and $(Y, N) = (0, 6)$ to Equation (17.1).

Exercise 17.3. By solving directly, show that $(H, L) = (0, 0)$ and $(4, 3)$ are equilibrium solutions to the following system of equations:

$$\frac{dH}{dt} = .3H - .1HL$$

$$\frac{dL}{dt} = .05HL - .2L$$

(17.14)

Exercise 17.4. A system of two differential equations has a Jacobian matrix at the equilibrium solution $(0, 0)$ as the following:

$$J_{(0,0)} = \begin{pmatrix} 0 & -1 \\ 1 & 0 \end{pmatrix}$$

(17.15)

What would be a system of differential equations that would produce that Jacobian matrix?

Exercise 17.5. Consider the following nonlinear system:

$$\frac{dx}{dt} = y - 1$$

$$\frac{dy}{dt} = x^2 - 1$$

(17.16)

a. Verify that this system has equilibrium solutions at $(-1, 1)$ and $(1, 1)$.
b. Determine the linear system associated with the tangent plane approximation at the equilibrium solution $(x, y) = (-1, 1)$ and $(1, 1)$ (two separate linear systems).
c. Construct the Jacobian matrix at the equilibrium solutions at $(-1, 1)$ and $(1, 1)$.
d. With the Jacobian matrix, visualize a phase plane at these equilbrium solutions to estimate stability of the equilibrium solution.

Exercise 17.6. (Inspired by Strogatz (2015)) Consider the following nonlinear system:

$$\frac{dx}{dt} = y - x$$

$$\frac{dy}{dt} = -y + \frac{5x^2}{4 + x^2}$$

(17.17)

a. Verify that the point $(x, y) = (1, 1)$ is an equilibrium solution.
b. Determine the linear system associated with the tangent plane approximation at the equilibrium solution $(x, y) = (1, 1)$.
c. Construct the Jacobian matrix at this equilibrium solution.
d. With the Jacobian matrix, visualize a phase plane at that equilbrium solution to estimate stability of the equilibrium solution.

Exercise 17.7. Consider the following system:

$$\frac{dx}{dt} = y^2$$

$$\frac{dy}{dt} = -x$$

(17.18)

a. Determine the equilibrium solution.
b. Visualize a phase plane of this system of differential equations.
c. Construct the Jacobian at the equilibrium solution.
d. Use the fact that $\frac{dy}{dt} \Big/ \frac{dx}{dt} = \frac{dy}{dx}$, which should yield a separable differential equation that will allow you to solve for a function $y(x)$. Plot several solutions of $y(x)$. How does that solution compare to the phase plane from the Jacobian matrix?

Exercise 17.8. The *Van der Pol Equation* is a second-order differential equation used to study radio circuits. In Chapter 16 you showed that the differential equations $x'' + \mu \cdot (x^2 - 1)x' + x = 0$, where μ is a parameter can be written as a system of equations:

$$\frac{dx}{dt} = y$$

$$\frac{dy}{dt} = -x - \mu(x^2 - 1)y$$

(17.19)

a. Determine the general Jacobian matrix $J_{(x,y)}$ for this system of equations.
b. The point $(0, 0)$ is an equilibrium solution. Evaluate the Jacobian matrix at the point $(0, 0)$. Your Jacobian matrix will depend on μ.
c. Evaluate your Jacobian matrix at the $(0, 0)$ equilibrium solution for different values of μ ranging from $-3, -1, 0, 1, 3$.
d. Make a phase plane for the Jacobian matrices at each of the values of μ.
e. Based on the phase planes that you generate, evaluate the stability of the equilibrium solution as μ changes.

Exercise 17.9. (Inspired by Logan and Wolesensky (2009)) A population of fish F has natural predators P. A model that describes this interaction is the following:

$$\frac{dF}{dt} = F - .3FP$$
$$\frac{dP}{dt} = .5FP - P$$
(17.20)

a. What are the equilibrium solutions for this differential equation?
b. Construct a Jacobian matrix for each of the equilibrium solutions.
c. Based on the phase plane from the Jacobian matrices, evaluate the stability of the equilibrium solutions.

Exercise 17.10. (Inspired by Pastor (2008)) The amount of nutrients (such as carbon) in soil organic matter is represented by N, whereas the amount of inorganic nutrients in soil is represented by I. A system of differential equations that describes the turnover of inorganic and organic nutrients is the following:

$$\frac{dN}{dt} = L + kdI - \mu NI - \delta N$$
$$\frac{dI}{dt} = \mu NI - kdI - \delta I$$
(17.21)

a. Verify that $N = \dfrac{L}{\delta}$, $I = 0$ and $N = \dfrac{kd + \delta}{\mu}$, $I = \dfrac{L\mu - \delta kd - \delta^2}{\mu\delta}$ are equilibrium solutions for this system.
b. Construct a Jacobian matrix for each of the equilibrium solutions.

Exercise 17.11. (Inspired by Logan and Wolesensky (2009) & Kermack, McKendrick, and Walker (1927)) A model for the spread of a disease where people recover is given by the following differential equation:

$$\frac{dS}{dt} = -\alpha SI$$
$$\frac{dI}{dt} = \alpha SI - \gamma I$$
(17.22)
$$\frac{dR}{dt} = \gamma I$$

Assume this population has $N = 1000$ people.

a. Determine the equilibrium solutions for this system of equations.
b. Construct the Jacobian for each of the equilibrium solutions.
c. Let $\alpha = 0.001$ and $\gamma = 0.2$. With the Jacobian matrix, generate the phase plane (using the equations for $\dfrac{dS}{dt}$ and $\dfrac{dI}{dt}$ only) for all of the equilibrium solutions and classify their stability.

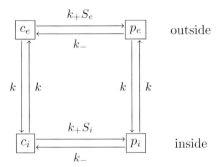

FIGURE 17.3 Glucose transporter reaction schemes.

Exercise 17.12. (Inspired by J. Keener and Sneyd (2009)) The chemical glucose is transported across the cell membrane using carrier proteins. These proteins can have different states (open or closed) that can be bound to a glucose substrate. The schematic for this reaction is shown in Figure 17.3. The system of differential equations describing this reaction is:

$$\frac{dp_i}{dt} = kp_e - kp_i + k_+s_ic_i - k_ip_i$$

$$\frac{dp_e}{dt} = kp_i - kp_e + k_+s_ec_e - k_-p_e$$

$$\frac{dc_i}{dt} = kc_e - kc_i + k_-p_i - k_+s_ic_i \qquad (17.23)$$

$$\frac{dc_e}{dt} = kc_i - kc_e + k_-p_e - k_+s_ec_e$$

a. We can reduce this to a system of three equations. First show that $\dfrac{dp_i}{dt} + \dfrac{dp_e}{dt} + \dfrac{dc_i}{dt} + \dfrac{dc_e}{dt} = 0$. Given that $p_i + p_e + c_i + c_e = C_0$, where C_0 is constant, use this equation to eliminate p_i and write down a system of three equations.

b. Determine the equilibrium solutions for this new system of three equations.

c. Construct the Jacobian matrix for each of these equilibrium solutions.

18

What are Eigenvalues?

18.1   Introduction

This chapter focuses on developing a tool to understand the stability of an equilibrium solution. This tool is determining eigenvalues and eigenvectors. We connect eigenvectors and eigenvalues back to straight-line solutions introduced in Chapter 15. You will see how eigenvalues are determined is through solving a polynomial equation. Finally we investigate how the values of the eigenvalues are reflected in the directions of the arrows in the phase plane. There is a lot here to unpack, so let's get started!

18.2   Straight line solutions

Consider this following linear system of equations:

$$
\begin{aligned}
\frac{dx}{dt} &= 2x - y \\
\frac{dy}{dt} &= 2x + 5
\end{aligned}
\tag{18.1}
$$

In Chapter 15 we identified two straight line solutions:

- Solution 1: $\vec{s}_1(t) = e^{4t} \begin{pmatrix} 1 \\ -2 \end{pmatrix}$
- Solution 2: $\vec{s}_2(t) = e^{3t} \begin{pmatrix} -1 \\ 1 \end{pmatrix}$

Let's verify that Solution 2 is indeed a solution to this linear system. First we will take the derivative of Solution 2:

$$
\frac{d}{dt}(\vec{s}_2(t)) = 3e^{3t} \begin{pmatrix} -1 \\ 1 \end{pmatrix} = \begin{pmatrix} -3e^{3t} \\ 3e^{3t} \end{pmatrix}
\tag{18.2}
$$

Let's compare this solution to the right hand side of the differential equation, recognizing that the x component of $\vec{s}_2(t)$ is $-e^{3t}$ and the y component of $\vec{s}_2(t)$ is e^{3t}:

$$2x - y = -2e^{3t} - e^{3t} = -3e^{3t}$$
$$2x + 5y = -2e^{3t} + 5e^{3t} = 3e^{3t}$$

(18.3)

So, indeed $\vec{s}_2(t)$ *is* a solution to the differential equation. However something interesting is occurring. Notice how $\dfrac{d}{dt}(\vec{s}_2(t))$ equals $3\vec{s}_2(t)$, which was the same as the right hand side of the differential equation.

While we wrote the right hand side of Equation (18.1) component by component, we could also write it as $A\vec{x}$, where $A = \begin{pmatrix} 2 & -1 \\ 2 & 5 \end{pmatrix}$. Because we verified $\vec{s}_2(t)$ was a solution to the differential equation, we could also have said that $A\vec{s}_2(t) = 3\vec{s}_2(t)$.

So we have two interesting facts here:

- A straight line solution to a system of linear differential equations $(\dfrac{d\vec{x}}{dt} = A\vec{x})$ has the form $\vec{s}(t) = c_1 e^{\lambda t} \vec{v}$, where c_1 is a constant and $\vec{v}$ a constant vector.
- Differentiating $\vec{s}(t)$ yields $\dfrac{d}{dt}(\vec{s}(t)) = \lambda \vec{s}(t)$.
- Consequently $\lambda \vec{s}(t) = A\vec{s}(t)$ in order for $\vec{s}(t)$ to be a solution.

All of these facts (in particular $\lambda \vec{s}(t) = A\vec{s}(t)$) set up an interesting equation: $\lambda c_1 e^{\lambda t} \vec{v} = c_1 e^{\lambda t} A\vec{v}$. Applying concepts from linear algebra, in order for the solution $\vec{s}(t)$ to be consistent, $A\vec{v} - \lambda I \vec{v} = \vec{0}$, where $\vec{0}$ is a vector of all zeros and I is called the *identity matrix*, or a square matrix with ones along the diagonal and zero everywhere else. The goal is to find a λ and $\vec{v}$ consistent with this equation.

Let's apply some terminology here. For these special straight line solutions, we give a particular name to $\vec{v}$ - we call it the *eigenvector*. The name we give to λ is the *eigenvalue*. (Eigen means *own* in German - get it?)

So how do we *determine* an eigenvalue or eigenvector? We do this by first determining the eigenvalues λ. This is done by solving the equation $\det(A - \lambda I) = 0$ for λ, where $\det(M)$ is the determinant. Once the eigenvalues are found, we then compute the eigenvectors by solving the equation $A\vec{v} - \lambda \vec{v} = \vec{0}$.

Let's take a time out. I recognize that we are starting to get deeper into linear algebra which may be some unfamiliar concepts. However we will just highlight key results that we will need - so hopefully that will give you a leg up when you study linear algebra - it is a great topic! Let's get to work.

18.3   Computing eigenvalues and eigenvectors

Let's dig into understanding the equation $\det(A - \lambda I) = 0$ for a two-linear system of differential equations. In this case, A is the matrix $\begin{pmatrix} a & b \\ c & d \end{pmatrix}$, for which then $A - \lambda I$ is the following matrix:

$$A - \lambda I = \begin{pmatrix} a - \lambda & b \\ c & d - \lambda \end{pmatrix} \tag{18.4}$$

The determinant of a 2×2 matrix is formed by the product of the diagonal entries less the product of the off-diagonal entries. For Equation (18.4), $\det(A - \lambda I) = 0$ is the equation $(a - \lambda)(d - \lambda) - bc = 0$. Our goal is to solve this equation for λ, which are the eigenvalues for this system.

Example 18.1. Compute the eigenvalues for the matrix $A = \begin{pmatrix} -1 & 1 \\ 0 & 3 \end{pmatrix}$.

Solution. The matrix $A - \lambda I$ is $A - \lambda I = \begin{pmatrix} -1 - \lambda & 1 \\ 0 & 3 - \lambda \end{pmatrix}$. So we have:

$$\det(A - \lambda I) = (-1 - \lambda)(3 - \lambda) - 0 = 0 \tag{18.5}$$

Solving the equation $(-1 - \lambda)(3 - \lambda) = 0$ yields two eigenvalues: $\lambda = -1$ or $\lambda = 3$.

More generally the equation $\det(A - \lambda I)$ yields a polynomial equation in λ. We call this equation the *characteristic polynomial* and denote it by $f(\lambda)$. In the case of a two-dimensional system of equations, $f(\lambda)$ will be a quadratic equation (see Exercise 18.9).

Once we have determined the eigenvalues, we next compute the eigenvectors associated with each eigenvalue. Remember that an eigenvector is a vector $\vec{v}$ consistent with $A\vec{v} = \lambda \vec{v}$ or $A\vec{v} - \lambda \vec{v} = \vec{0}$. How we do this is through algebra, as is done in the following example:

Example 18.2. Compute the eigenvectors for the matrix $A = \begin{pmatrix} -1 & 1 \\ 0 & 3 \end{pmatrix}$ from Example 18.1.

Solution. First for general λ, consider the expression $A\vec{v} - \lambda \vec{v} = \vec{0}$, where $\vec{v} = \begin{pmatrix} v_1 \\ v_2 \end{pmatrix}$:

$$A\vec{v} - \lambda \vec{v} = \vec{0} \rightarrow \begin{pmatrix} -v_1 + v_2 - \lambda v_1 = 0 \\ 3v_2 - \lambda v_2 = 0 \end{pmatrix} \tag{18.6}$$

We use the two expressions $(-v_1 + v_2 - \lambda v_1 = 0$ and $3v_2 - \lambda v_2 = 0)$ in Equation (18.6) to determine the eigenvector $\vec{v}$. We need to consider both eigenvalues $(\lambda = -1$ and $\lambda = 3)$ separately to yield two different eigenvectors:

- **Case 1** $\lambda = -1$: The first expression in Equation (18.6) yields $-v_1 + v_2 + v_1 = 0$, or $v_2 = 0$ after simplifying. For the second expression we have $3v_2 + v_2 = 0 \rightarrow 4v_2 = 0$, so that tells us again that $v_2 = 0$. Notice we've determined a value for the second component v_2, but not v_1. In this case, we say that the first component to the vector $\vec{v}$ is a *free parameter.*, so the eigenvector can be expressed as $\begin{pmatrix} c_1 \\ 0 \end{pmatrix}$, where c_1 is the free parameter. Another way to express this eigenvector is $c_1 u_1$, with $u_1 = \begin{pmatrix} 1 \\ 0 \end{pmatrix}$. The eigenvector in this case is $\begin{pmatrix} 1 \\ 0 \end{pmatrix}$. (We generally write eigenvectors without the arbitrary constants.) This particular straight line solution is $s_1(t) = c_1 e^{-t} u_1$, where c_1 is a free variable.

- **Case 2** $\lambda = 3$: For the second equation we have $3v_2 - 3v_2 = 0$, which is always true. However in the first equation we have $-v_1 + v_2 - 3v_1 = 0$, or $v_2 = 4v_1$. In this case, v_1 can be a free parameter; however, v_2 will have to be four times the value of v_1. Hence, this particular straight line solution is $s_2(t) = e^{3t} \begin{pmatrix} c_2 \\ 4c_2 \end{pmatrix}$, or also as $s_2(t) = c_2 e^{3t} \vec{u}_2$, with $\vec{u}_2 = \begin{pmatrix} 1 \\ 4 \end{pmatrix}$. The eigenvector in this case is $\begin{pmatrix} 1 \\ 4 \end{pmatrix}$.

Notice that in both of our cases we had a free variable (c_1 or c_2), which are also constants in our final solution for the differential equation.

Once we have computed the eigenvalues and eigenvectors, we are now ready to express the most general solution for a system of differential equations. For a two-dimensional system of linear differential equations $(\frac{d}{dt}\vec{x} = A\vec{x})$, the most general solution is given by Equation (18.7):

$$\vec{x}(t) = c_1 e^{\lambda_1 t} \vec{v}_1 + c_2 e^{\lambda_2 t} \vec{v}_2 \tag{18.7}$$

Example 18.3. What is the solution to the differential equation $\frac{d}{dt}\vec{x} = \begin{pmatrix} -1 & 1 \\ 0 & 3 \end{pmatrix} \vec{x}$?

Solution. Since we have already computed the eigenvalues and eigenvectors, our most general solution for this differential equation is:

$$\vec{x} = c_1 e^{-t} \begin{pmatrix} 1 \\ 0 \end{pmatrix} + c_2 e^{3t} \begin{pmatrix} 1 \\ 4 \end{pmatrix},$$

with c_1 and c_2 defined as constants.

18.3.1   Computing eigenvalues with demodelr

While computing eigenvalues and eigenvectors is a good algebraic exercise, we can also program this in R using the function `eigenvalues` from the `demodelr` package. The syntax works where $A = \begin{pmatrix} a & b \\ c & d \end{pmatrix}$ is entered in as `eigenvalues(a,b,c,d,matrix_rows)` where `matrix_rows` is the number of rows.[1] What gets returned from the function will be the eigenvalues and eigenvectors for any square matrix.

Let's compute the eigenvalues for the matrix $\begin{pmatrix} -1 & 1 \\ 0 & 3 \end{pmatrix}$:

```
# For a two-dimensional equation the code assumes
# the default is a 2 by 2 matrix.

demodelr::eigenvalues(matrix_entries = c(-1, 1, 0, 3),
            matrix_rows = 2)

## eigen() decomposition
## $values
## [1]   3 -1
##
## $vectors
##             X1 X2
## 1 0.2425356   1
## 2 0.9701425   0
```

Notice that the eigenvalues and the eigenvectors get returned with the `eigenvalues` function. How you read the output for the eigenvector is that X1 is the eigenvector associated with the first eigenvalue ($\lambda = 3$) and X2 is the eigenvector associated with the second eigenvalue ($\lambda = -1$). The eigenvector associated with $\lambda = 3$ is a little different from what we computed - R will *normalize* the vector, which means that its total length will be one.[2] For example $\vec{v}_2 = \begin{pmatrix} 1 \\ 4 \end{pmatrix}$, so $||\vec{v}|| = \sqrt{5}$, and the normalized vector is $\begin{pmatrix} 1/\sqrt{5} \\ 4/\sqrt{5} \end{pmatrix}$, which you can verify is the same as the reported eigenvector from the `eigenvalues` function.

---

[1]If you have a 2 by 2 matrix, you can leave out `matrix_rows` (so just `eigenvalues(a,b,c,d)`) as the default is a 2 by 2 matrix.

[2]The length of a vector $\vec{v}$ is denoted as $||\vec{v}||$ and is computed the following way: $||\vec{v}|| = \sqrt{v_1^2 + v_2^2 + ... + v_n^2}$. We normalize a vector to a length of 1 by dividing each component by its length.

18.4   What do eigenvalues tell us?

Here the focus of the chapter changes a little bit. Now we focus on understanding how the phase plane for a differential equation gives clues about the stability for an equilibrium solution. This is intentional: once we have found the eigenvalues, determining eigenvectors can seem rather mundane at times (and perhaps heavy on the algebra). Studying the eigenvalues helps us understand the qualitative nature of the solution to a differential equation. Let's think about the characteristic equation $f(\lambda)$ for a two-dimensional system of differential equations:

$$
\begin{aligned}
f(\lambda) &= \det(A - \lambda I) \\
&= (a - \lambda)(d - \lambda) - bc \\
&= \lambda^2 - (a + d)\lambda + ad - bc
\end{aligned}
\tag{18.8}
$$

Notice how $f(\lambda)$ is a quadratic equation. You may recall that quadratic equations have zero, one, or two distinct solutions. If there are no solutions, we say the solutions are imaginary (more on that later). Also the signs of solutions may be positive or negative. There are so many different combinations! What types of phase planes do all those different types of eigenvalues produce? The following examine representative examples of all the possible eigenvalues you may obtain for a two-dimensional linear system of differential equations (which can be generalized to higher-dimensional systems).

18.4.1   Sources: all eigenvalues positive

Consider the differential equation in Equation (18.9):

$$
\begin{pmatrix} x' \\ y' \end{pmatrix} = \begin{pmatrix} 2 & 0 \\ 1 & 1 \end{pmatrix} \begin{pmatrix} x \\ y \end{pmatrix}
\tag{18.9}
$$

Computing the eigenvalues for Equation (18.9) shows that they are both positive:

```
eigenvalues(c(2, 0, 1, 1))
```

```
## eigen() decomposition
## $values
## [1] 2 1
##
## $vectors
##           X1 X2
## 1 0.7071068  0
## 2 0.7071068  1
```

The phase plane for this matrix A is shown in Figure 18.1:

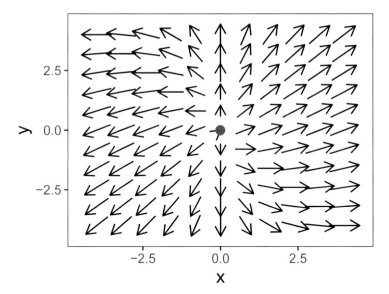

FIGURE 18.1 Phase plane for Equation (18.9), which shows the equilibrium solution is a source (also known as an unstable node).

Notice how the phase plane in Figure 18.1 has all the arrows pointing from the origin. In this case we call the origin equilibrium solution a *source*, or also an *unstable node*. Plotting the components of $\vec{x}(t)$ as functions of t would show the dependent values increase exponentially as time increases.

18.4.2 Sinks: all eigenvalues negative

Consider the differential equation in Equation (18.10), which is a *slight* modification from Equation (18.9):

$$\begin{pmatrix} x' \\ y' \end{pmatrix} = \begin{pmatrix} -2 & 0 \\ 1 & -1 \end{pmatrix} \begin{pmatrix} x \\ y \end{pmatrix} \tag{18.10}$$

The eigenvalues for Equation (18.10) are both negative (verify this on your own). The resulting phase plane for Equation (18.10) then has all the arrows pointing towards the origin, shown in Figure 18.2.

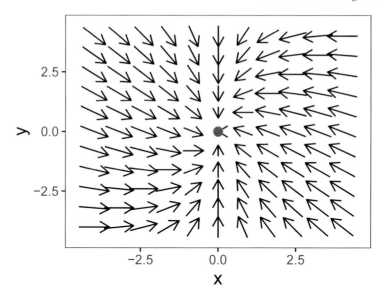

FIGURE 18.2 Phase plane for Equation (18.10), which shows the equilibrium solution is a sink (also known as a stable node).

Based on the phase plane in Figure 18.2, solutions to Equation (18.10) would asymptotically approach the origin. We say the equilibrium solution is a *sink*, also known as a *stable node*.

18.4.3 Saddle nodes: one positive and one negative eigenvalue

Consider the differential equation in Equation (18.11):

$$\begin{pmatrix} x' \\ y' \end{pmatrix} = \begin{pmatrix} 3 & -2 \\ 1 & -1 \end{pmatrix} \begin{pmatrix} x \\ y \end{pmatrix} \tag{18.11}$$

Equation (18.11) has $\lambda_1 = 2.414$ and $\lambda_2 = -0.414$ (verify this on your own). Because the differential equation has one positive and one negative eigenvalue the equilibrium solution the phase plane for this differential equation looks a little different, as is shown in Figure 18.3:

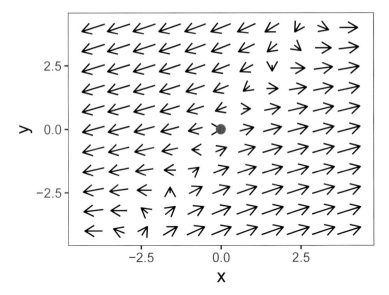

FIGURE 18.3 Phase plane for Equation (18.11), which shows the equilibrium solution is a saddle node.

This equilibrium solution is called a *saddle node*. From the horizontal direction, the arrows point *away* from the origin, but in the vertical direction the arrows point *towards* the origin. This behavior is caused by the opposing signs of the eigenvalues - one part of the solution in Equation (18.7) (the one associated with the negative eigenvalue) decays asymptotically to zero. The other positive eigenvalue is associated with the asymptotically unstable, giving the solution trajectory the shape of a saddle.

18.4.4 Spirals: imaginary eigenvalues

Consider the differential equation in Equation (18.12):

$$\begin{pmatrix} x' \\ y' \end{pmatrix} = \begin{pmatrix} -3 & -8 \\ 4 & -6 \end{pmatrix} \begin{pmatrix} x \\ y \end{pmatrix} \tag{18.12}$$

There are two eigenvalues to this system: $\lambda = -4.5 + 5.45i$ and $\lambda = -4.5 - 5.45i$ (you can confirm this on your own). In this case the i means the eigenvalues are imaginary. Notice how the eigenvalues are similar, but the signs on the second term differ. We say the eigenvalues are *complex conjugates* of each other, and write them in the form $\lambda = \alpha \pm \beta i$. In this example $\alpha = -4.5$ and $\beta = 5.45$. Figure 18.4 shows the phase plane for this system.

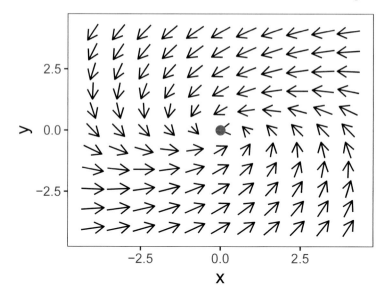

FIGURE 18.4 Phase plane for Equation (18.12), which shows the equilibrium solution is a spiral sink.

The phase plane in Figure 18.4 has some spiraling motion to it. Why does that occur? Imaginary eigenvalues can occur when the characteristic equation $\det(A - \lambda I) = 0$ has imaginary solutions. More generally, we say $\lambda = \alpha \pm \beta i$. Because the eigenvalues are complex, we would also expect the eigenvectors to be complex as well (i.e. $\vec{v} \pm i\vec{w}$). Don't let the term *imaginary* fool you: by using properties from complex analysis it can be shown that when eigenvalues are imaginary, the template for the solution is given in Equation (18.13):

$$\vec{x}(t) = c_1 e^{\alpha t}(\vec{w}\cos(\beta t) - \vec{v}\sin(\beta t)) + c_2 e^{\alpha t}(\vec{w}\cos(\beta t) + \vec{v}\sin(\beta t)) \quad (18.13)$$

The trigonometric terms in Equation (18.13) suggest that the solution has some periodic behavior if we plot the components of $\vec{x}(t)$ as functions of t. But when we plot the solution in the xy plane that periodic behavior gets translated to spiraling motion in Figure 18.4. When $\alpha < 0$ we say the equilibrium solution is a *spiral sink* because the exponential terms in Equation (18.13) decay asymptotically to zero.

As you would expect when $\alpha > 0$ we classify a phase plane as a *spiral source* (shown in Equation (18.14) and Figure 18.5).

$$\begin{pmatrix} x' \\ y' \end{pmatrix} = \begin{pmatrix} 4 & -5 \\ 3 & 2 \end{pmatrix} \begin{pmatrix} x \\ y \end{pmatrix} \quad (18.14)$$

().

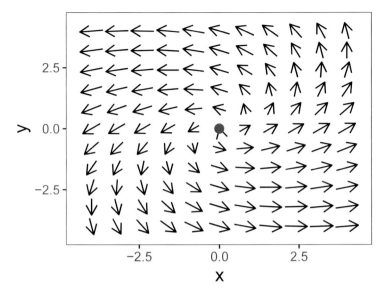

FIGURE 18.5 Phase plane for Equation (18.14), which shows the equilibrium solution is a spiral source.

The final case for imaginary eigenvalues is when $\alpha = 0$, which is termed a *center*. As an example, let's examine the phase plane for the system in Equation (18.15):

$$\begin{pmatrix} x' \\ y' \end{pmatrix} = \begin{pmatrix} 0 & -1 \\ 1 & 0 \end{pmatrix} \begin{pmatrix} x \\ y \end{pmatrix} \tag{18.15}$$

$$f(\lambda) = \lambda^2 - \text{tr}(A)\lambda + \det(A) \tag{19.4}$$

Example 19.1. Determine the characteristic polynomial $f(\lambda)$ for the system $\vec{x}' = Ax$ where $A = \begin{pmatrix} -1 & 1 \\ 0 & 3 \end{pmatrix}$. Solve for the eigenvalues to classify the stability of the equilibrium solution.

Solution. We can see that $\det(A) = -1(3) - 0(1) = -3$ and $\text{tr}(A) = 2$, so our characteristic equation is $\lambda^2 - 2\lambda - 3$. If we solve $\lambda^2 - 2\lambda - 3 = 0$ we have $(\lambda - 3)(\lambda + 1) = 0$, so our eigenvalues are $\lambda = 3$ and $\lambda = -1$. Since one eigenvalue is positive and the other one is negative, the equilibrium solution is a saddle node.

As shown in Example 19.1, Equation (19.4) may be a computationally easier way to determine eigenvalues. Here is *another* way to think about eigenvalues. Let's say we have two eigenvalues λ_1 and λ_2, so $f(\lambda_1) = 0$ and $f(\lambda_2) = 0$. However we can also examine Equation (19.5) in terms of the roots of $f(\lambda)$, denoted as λ_1 and λ_2. We make no assumptions about whether λ_1 or λ_2 are real, imaginary, or equal in value. However, since they solve the equation $f(\lambda) = 0$, this also means that $(\lambda - \lambda_1)(\lambda - \lambda_2) = 0$. If we multiply out this equation we have $\lambda^2 - (\lambda_1 + \lambda_2)\lambda + \lambda_1\lambda_2 = 0$. If we compare this equation with Equation (19.4) we have:

$$\begin{aligned} f(\lambda) &= \lambda^2 - (\lambda_1 + \lambda_2)\lambda + \lambda_1\lambda_2 \\ &= \lambda^2 - \text{tr}(A)\lambda + \det(A) \end{aligned} \tag{19.5}$$

Equation (19.5) allows us to identify that $\text{tr}(A) = \lambda_1 + \lambda_2$ and $\det(A) = \lambda_1\lambda_2$, or the trace of A is the *sum* of the two eigenvaules and the determinant of A is the *product* of the eigenvalues. Let's explore this a little more.

19.2   Stability with the trace and determinant

The equality in Equation (19.5) uncovers some neat relationships - in particular $\text{tr}(A) = (\lambda_1 + \lambda_2)$ and $\det(A) = \lambda_1 + \lambda_2$. Table 19.1 synthesizes all these relationships to provide an alternative pathway to understand stability of an equilibrium solution with the trace and determinant:

TABLE 19.1 Comparison of the stability of an equilibrium solution in relation to the signs of an eigenvalue, the trace of the matrix A, or the determinant of the matrix A.

Sign of λ_1	Sign of λ_2	Tendency of equilibrium solution	Sign of $\text{tr}(A) = \lambda_1 + \lambda_2$	Sign of $\det(A) = \lambda_1 \cdot \lambda_2$
Positive	Positive	Source	Positive	Positive
Negative	Negative	Sink	Negative	Positive
Positive	Negative	Saddle	?	Negative
Negative	Positive	Saddle	?	Negative

For the moment we will only consider real non-zero values of the eigenvalues - more specialized cases will occur later. But examining the above table carefully:

- If $\det(A)$ is *negative*, then the equilibrium solution is a *saddle*.
- If $\det(A)$ is *positive* and $\text{tr}(A)$ is *negative*, then the equilibrium solution is a *sink*.
- If $\det(A)$ and $\text{tr}(A)$ are both *positive*, then the equilibrium solution is a *source*.

Example 19.2. Use the trace and determinant relationships to classify the stability of the equilibrium solution for the linear system $\vec{x}' = A\vec{x}$ where $A = \begin{pmatrix} -1 & 1 \\ 0 & 3 \end{pmatrix}$.

Solution. We can see that $\det(A) = -1(3) - 0(1) = -3$ and $\text{tr}(A) = 2$. Since the determinant is negative, the equilibrium solution must be a saddle node.

Knowing the relationships between the trace and determinant for a two-dimensional system of equations is a pretty quick and easy way to investigate stability of equilibrium solutions!

Another way to graphically represent the stability of solutions is with the *trace-determinant plane* (shown in Figure 19.1), with $\text{tr}(A)$ on the horizontal axis and $\det(A)$ on the vertical axis:

e. Given these constraints, what would the phase plane for this system be?
f. Create a linear two-dimensional system where $\text{tr}(A) = 0$ and $\det(A) > 0$. Show your system and the phase plane.

20

Bifurcation

In this chapter we will use *bifurcation* to examine how the stability of an equilibrium solution changes as the value of a parameter changes. This is a great topic of study that (by necessity) requires you to think of stability of an equilibrium solution on multiple levels. You are up for the challenge; let's get started!

20.1 A series of equations

Consider the differential equation $x' = 1 - x^2$. This differential equation has an equilibrium solution at $x = \pm 1$. To classify the stability of the equilibrium solutions we apply the following test for stability of an equilibrium solution that we developed in Chapter 5:

- If $f'(y^*) > 0$ at an equilibrium solution, the equilibrium solution $y = y^*$ will be unstable.
- If $f'(y^*) < 0$ at an equilibrium solution, the equilibrium solution $y = y^*$ will be stable.
- If $f'(y^*) = 0$, we cannot conclude anything about the stability of $y = y^*$.

Applying this test, we know $f(x) = 1 - x^2$ and $f'(x) = -2x$. Since $f'(1) = -2$ and $f'(-1) = 2$, then the respective equilibrium solution $x = 1$ is stable and the equilibrium solution at $x = -1$ is unstable.

Let's modify and extend this example further. Consider two more differential equations:

- $x' = -1 - x^2$: This differential equation does not have any equilibrium solutions, so we do not need to apply the stability test.
- $x' = -x^2$: This differential equation has an equilibrium solution at $x = 0$; the stability test cannot apply because $f' = -2x$ and $f'(0) = 0$. The general solution to this differential equation is $x(t) = \dfrac{1}{t + C}$ (Exercise 20.2), which apart from the vertical asymptote at $t = -C$ is always decreasing for $t > 0$. So the equilibrium solution at $x = 0$ is not stable.

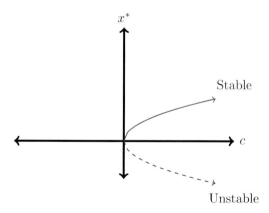

FIGURE 20.2 A saddle node bifurcation for the differential equation $x' = c - x^2$.

Let's talk about Figure 20.2. The graph represents the value of the equilibrium solution (x^*, vertical axis) as a function of the parameter c (horizontal axis). Since equilibrium solutions are characterized by $x^* = \pm\sqrt{c}$ we have the "sideways parabola," traced in blue in Figure 20.2. When $c < 0$, there is no equilibrium solution, (so nothing is plotted in the second and third quadrants of Figure 20.2). The difference between the solid and dashed lines in Figure 20.2 is used to distinguish between a stable equilibrium solution ($x^* = \sqrt{c}$ when $c > 0$) and an unstable equilibrium solution ($x^* = -\sqrt{c}$ when $c > 0$). It is so cool that *all* the information about the equilibrium solution and its stability is contained in Figure 20.2!

The bifurcation structure of $x' = c - x^2$ is called a *saddle-node* bifurcation. To give another context, it might be helpful to think of this c like a tuning knob. As $c > 0$ we will always have two different equilibrium solutions that are symmetrical based on the value of c. The positive equilibrium solution will be stable, the other unstable. As c approaches zero these equilibrium solutions will collapse into each other. If c is negative, the equilibrium solution disappears.

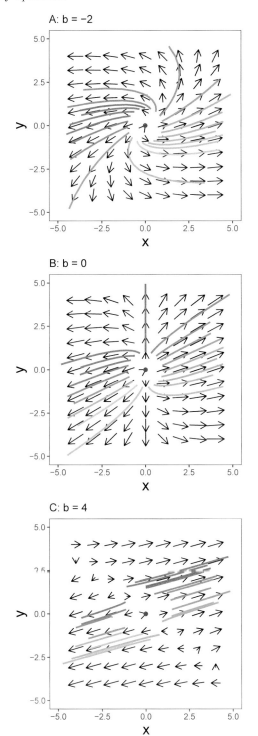

FIGURE 20.3 Phase planes for Equation (20.1) for different values of b.

20.2   Bifurcations with systems of equations

We can also examine how bifurcation plays a role with systems of differential equations.

As another example, let's determine the behavior of solutions near the origin for the system of equations:

$$\frac{d\vec{x}}{dt} = \begin{pmatrix} 3 & b \\ 1 & 1 \end{pmatrix} \vec{x}. \tag{20.1}$$

This equation has one free parameter b that we will analyze using the trace determinant conditions developed in Chapter 19. Let's call the matrix A, so the $\text{tr}(A) = 4$ and $\det(A) = 3 - b$. Since the trace is always positive either the equilibrium solution will be a saddle if $\det(A) < 0$, or when $3 < b$. We have a spiral source when $\det(A) > 0$ (this means $3 > b$) and $\det(A) > (\text{tr}(A))^2/4$, or when $3 - b > 4$, which leads to $b < -1$. Figure 20.3A-C displays the phase planes for different values of b along with sample solution curves.

To summarize, Equation (20.1) has the following dynamics depending on the value of b:

- When $b < -1$, the equilibrium solution will be a spiral source.
- When $-1 < b < 3$, the equilibrium solution will be a source.
- When $3 < b$, the equilibrium solution will a saddle.

Another approach to analyzing Equation (20.1) is to compute the eigenvalues directly, which in this case are $\lambda_{1,2}(b) = 2 \pm \sqrt{b+1}$. Creating a plot of the eigenvalues (Figure 20.4) can also help explain the bifurcation structure. When $b < -1$, the eigenvalues are imaginary, with $\text{Re}(\lambda_{1,2}(b) = 2) = 2$, so the equilibrium solution is a spiral source. When $-1 < b < 3$, both eigenvalues are positive, so the equilibrium solution is a source. Finally, when $3 < b$, one eigenvalue is positive and the other is negative, confirming our analyses with the trace-determinant plane.

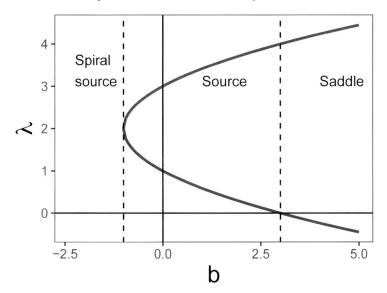

FIGURE 20.4 Bifurcation diagram for Equation (20.1). The vertical axis shows the value of the eigenvalues λ (red and blue curves) as a function of the parameter b. The annotations represent the stability of the original equilibrium solution.

The benefit of a bifurcation diagram is to provide a complete understanding of the dynamics of the system *as a function of the parameters*. In this chapter we examined *one-parameter* bifurcations (for example we looked the stability of the equilibrium solution as it depends on c or b), but bifurcations can also be extended further to two parameter bifurcation families, applying similar methods. In general the methods are similar to what we have done.

20.3 Functions as equilibrium solutions: limit cycles

In the previous examples the stability of an equilibrium solution changed depending on the value of a parameter. Typically equilibrium solutions are a single point in the phase plane. Another way we can represent an equilibrium solution is with a *function*. As an example, consider the following highly nonlinear system in Equation (20.2):

$$\frac{dx}{dt} = -y - x(x^2 + y^2 - 1)$$
$$\frac{dy}{dt} = x - y(x^2 + y^2 - 1)$$

$$(20.2)$$

The phase plane for Equation (20.2) is shown in Figure 20.5. You can verify
that Equation (20.2) has an equilibrium solution at the point $x = 0$, $y = 0$.
However Figure 20.5 suggests there might be other equilibrium solutions when
various solution curves are plotted with the phase plane.

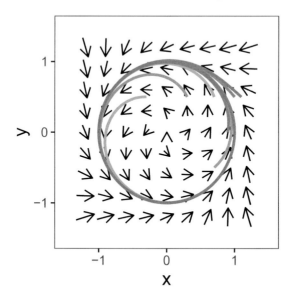

FIGURE 20.5 Phase plane for Equation (20.2) with different solution curves.
Notice the equilibrium solution described by the equation $x^2 + y^2 = 1$.

What is interesting in Figure 20.5 is that the solution tends towards a circle of
radius 1 (or the equation $x^2 + y^2 = 1$). This is an example of an equilibrium
solution that is a *curve* rather than a specific point. We can describe the phase
plane with a new variable X that represents the distance from the origin (a
radius r) by transforming this system from x and y to a single new variable X
(see Exercise 20.11).

$$\frac{dX}{dt} = -X(X - 1) \text{ where } X = r^2 \qquad (20.3)$$

How Equation (20.2) transforms to Equation (20.3) is by applying a polar
coordinate transformation to this system. With stability analysis for Equation
(20.3) we can show that the equilibrium solution $X = 0$ is unstable (meaning
the origin $x = 0$ and $y = 0$ is an unstable equilibrium solution) and the circle
of radius 1 is a stable equilibrium solution (which is the equation $x^2 + y^2 = 1$).
In this case we would say $r = 1$ is a *stable limit cycle*. You will study a similar
system in Exercises 20.11 and 20.12.

Equation (20.3) is an example of next steps with studying the qualitative
analysis of systems. We can extend out Equation (20.2) further to introduce a

parameter μ that, as μ changes, undergoes a bifurcation as μ increases. This is an example of a *Hopf bifurcation*.

20.4 Bifurcations as analysis tools

The most important part in studying bifurcations is analyzing examples. This chapter has several exercises where you will construct bifurcation diagrams for one- and two-dimensional systems of differential equations. As a reminder, constructing sample phase lines / phase planes before analyzing stability and the bifurcation structure is always helpful to build understanding.

Bifurcation analysis is a fascinating field of study that combines knowledge of differential equations, geometry, and other types of advanced mathematics. For further information, please see the texts by Strogatz (2015), Perko (2001), and Kuznetsov (2004).

20.5 Exercises

Exercise 20.1. Explain why $x' = 1 + x^2$ does not have any equilibrium solutions.

Exercise 20.2. Use separation of variables to verify that the general solution to $x' = -x^2$ is $x(t) = \dfrac{1}{t + C}$.

Exercise 20.3. Apply local linearization to classify stability of the following differential equations:

a. $\dfrac{dx}{dt} = x - x^2$

b. $\dfrac{dx}{dt} = -x^2$

c. $\dfrac{dx}{dt} = -x - x^2$

Exercise 20.4. Consider the differential equation $x' = cx - x^2$. What are equations that describe the dependence of the equilibrium solution on the value of c? Once you have that figured out, plot the bifurcation diagram, with the parameter c along the horizontal axis. This bifurcation is called a *transcritical* bifurcation.

Exercise 20.5. Consider the differential equation $x' = cx - x^3$. What are equations that describe the dependence of the equilibrium solution on the value of c? Once you have that figured out plot the bifurcation diagram, with the parameter c along the horizontal axis. This bifurcation is called a *pitchfork* bifurcation.

Exercise 20.6. Construct a bifurcation diagram for the differential equation $x' = c + x^2$

Exercise 20.7. Consider the differential equation $x' = x(x - 1)(b - x)$. The differential equation has equilibrium solutions at $x^* = 0$, $x^* = 1$, and $x^* = b$, where $b > 0$.

a. Use desmos or some other plotting software to investigate the effect of the number of roots as b increases from a value of 0.
b. Analyze the stability of each of these equilibrium solutions. (You may want to multiply out the right hand side of the differential equation.) Whether a given equilibrium solution is stable may depend on the value of b.
c. Construct a bifurcation diagram for all three solutions together, with b on the horizontal axis and the value of x^* on the vertical axis.

Exercise 20.8. Consider the system of differential equations:

$$\begin{pmatrix} x' \\ y' \end{pmatrix} = \begin{pmatrix} -x \\ cy - y^2 \end{pmatrix} \tag{20.4}$$

a. What are the equilibrium solutions for this (uncoupled) system of equations?
b. Evaluate stability of the equilibrium solutions as a function of the parameter c.
c. Construct a few representative phase planes to verify your analysis.

Exercise 20.9. Consider the following linear system of differential equations:

$$\frac{d}{dt}\vec{x} = \begin{pmatrix} 3 & b \\ b & 1 \end{pmatrix}\vec{x}. \tag{20.5}$$

a. Verify that the characteristic polynomial is $f(\lambda, b) = \lambda^2 - 4\lambda + (3 - b^2)$.
b. Solve $f(\lambda, b) = 0$ with the quadratic formula to obtain an expression for the eigenvalues as a function of b, that is $\lambda_{1,2}(b)$.
c. Using the eigenvalues, classify the stability of the equilibrium solution as b changes.
d. Generate a few representative phase planes to verify your analysis.
e. Create a plot similar to Figure 20.4 showing the bifurcation structure.

Exercise 20.10. Consider the linear system of differential equations:

$$\frac{dx}{dt} = cx - y$$
$$\frac{dy}{dt} = -x + cy$$
                                                                            (20.6)

a. Determine the characteristic polynomial ($f(\lambda, c)$) for this system of equations.
b. Solve $f(\lambda, c) = 0$ with the quadratic formula to obtain an expression for the eigenvalues as a function of c, that is $\lambda_{1,2}(c)$.
c. Using the eigenvalues, classify the stability of the equilibrium solution as c changes.
d. Generate a few representative phase planes to verify your analysis.
e. Create a plot similar to Figure 20.4 showing the bifurcation structure.

Exercise 20.11. Consider the following highly nonlinear system:

$$\frac{dx}{dt} = -y - x(x^2 + y^2 - 1)$$
$$\frac{dy}{dt} = x - y(x^2 + y^2 - 1)$$
                                                                            (20.7)

We are going to transform the system by defining new variables $x = r\cos\theta$ and $y = r\sin\theta$. Observe that $r^2 = x^2 + y^2$.

a. Consider the equation $r^2 = x^2 + y^2$, where r, x, and y are all functions of time. Apply implicit differentiation to determine a differential equation for $\frac{d(r^2)}{dt}$, expressed in terms of x, y, $\frac{dx}{dt}$ and $\frac{dy}{dt}$.

b. Multiply the above equations $\frac{dx}{dt}$ by $2x$ and $\frac{dy}{dt}$ by $2y$ on both sides of the equation. Then add the two equations together. You should get an expression for $\frac{d(r^2)}{dt}$ in terms of x and y.

c. Rewrite the equation for the right hand side of $\frac{d(r^2)}{dt}$ in terms of r^2.

d. Use your equation that you found to verify that

$$\frac{dX}{dt} = -2X(X - 1), \text{ where } X = r^2$$
                                                                            (20.8)

e. Verify that $X = 1$ is a stable node and $X = 0$ is unstable.

f. As discussed in this chapter, this system has a stable limit cycle. What quick and easy modification to our system could you do to the system to ensure that this is an unstable limit cycle? Justify your work.

Exercise 20.12. Construct a bifurcation diagram for $\dfrac{dX}{dt} = -2X(X - \mu),; \mu$ is a parameter. Explain how you can apply that result to understanding the bifurcation diagram of the system:

$$\frac{dx}{dt} = -y - x(x^2 + y^2 - \mu)$$
$$\frac{dy}{dt} = x - y(x^2 + y^2 - \mu) \tag{20.9}$$

This system is an example of a *Hopf bifurcation*.

Exercise 20.13. (Inspired by Logan and Wolesensky (2009)) The immune response to HIV can be described with differential equations. In the early stages (before the body is swamped by the HIV virions) the dynamics of the virus can be described by the following system of equations, where v is the virus load and x the immune response:

$$\frac{dv}{dt} = rv - pxv$$
$$\frac{dx}{dt} = cv - bx \tag{20.10}$$

You may assume that all parameters are positive.

a. Explain the various terms in this model and their biological meaning.
b. Determine the equilibrium solutions.
c. Evaluate the Jacobian for each of the equilibrium solutions.
d. Construct a bifurcation diagram for each of the equilibrium solutions.

Part IV

Stochastic Differential Equations

21

Stochastic Biological Systems

21.1   Introducing stochastic effects

Up to this point we have studied *deterministic* differential equations. We use the word deterministic because given an initial condition and parameters, the solution trajectory is known. In this part we are going to study *stochastic* differential equations or SDEs for short. A stochastic differential equation means that the differential equation is subject to random effects - either in the parameters (which may cause a change in the stability for a time) or in the variables themselves.

Stochastic differential equations can be studied using computational approaches. This part will give you an introduction to SDEs with some focus on solution techniques, which I hope you will be able to apply in other contexts relevant to you. Understanding how to model SDEs requires learning some new mathematics and approaches to numerical simulation. Let's get started!

21.2   A discrete dynamical system

Let's focus on an example that involves discrete dynamical systems. Moose are large animals (part of the deer family), weighing 1000 pounds, that can be found in Northern Minnesota[1]. The moose population was 8000 in the early 2000s, but recent surveys[2] show the population is maybe stabilized at 3000.

A starting model that describes their population dynamics is the discrete dynamical system in Equation (21.1):

$$M_{t+1} = M_t + b \cdot M_t - d \cdot M_t, \tag{21.1}$$

where M_t is the population of the moose in year t, and b the birth rate and d the death rate. Equation (21.1) can be reduced down to $M_{t+1} = rM_t$ where

---

[1] https://www.dnr.state.mn.us/mammals/moose.html
[2] https://files.dnr.state.mn.us/wildlife/moose/moosesurvey.pdf

$r = 1 + b - d$ is the net birth/death rate. This model states that the population of moose in the next year is proportional to the current population.

Equation (21.1) is a little bit different from a continuous dynamical system, but can be simulated pretty easily by defining a function.

```
M0 <- 3000 # Initial population of moose
N <- 5 # Number of years we simulate

moose <- function(r) {
  out_moose <- array(M0, dim = N+1)
  for (i in 1:N) {
    out_moose[i + 1] <- r * out_moose[i]
  }
  return(out_moose)
}
```

Notice how the function moose returns the current population of moose after N years with the net birth rate r. Let's take a look at the results for different values of r (Figure 21.1).

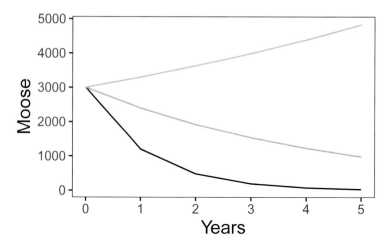

FIGURE 21.1 Simulation of the moose population with different birth rates.

Notice how for some values of r the population starts to decline, stay the same, or increase. To analyze Equation (21.1), just like with a continuous differential equation we want to look for solutions that are in steady state, or ones where the population is staying the same. In other words this means that $M_{t+1} = M_t$, or $M_t = rM_t$. If we simplify this expression this means that $M_t - rM_t = 0$, or

$(1 - r)M_t = 0$. Assuming that M_t is not equal to zero, then this equation is consistent only when $r = 1$. This makes sense: we know $r = 1 - b - d$, so the only way this can be one is if $b = d$, or the births balance out the deaths.

Okay, so we found our equilibrium solution. The next goal is to determine the general solution to Equation (21.1). In Chapter 7 for continuous differential equations, a starting point for a general solution was an exponential function. For discrete dynamical systems we will also assume a general solution is exponential, but this time we represent the solution as $M_t = M_0 \cdot v^t$, which is an exponential equation. The parameter M_0 is the initial population of moose (here it equals 3000). Now let's determine v in Equation (21.1):

$$M_{t+1} = rM_t \rightarrow 3000 \cdot v^{t+1} = r \cdot 3000 \cdot v^t \tag{21.2}$$

Our goal is to figure out a value for v that is consistent with this expression. Just like we did with continuous differential equations we can arrange the following equation, using the fact that $v^{t+1} = v^t \cdot v$:

$$3000 v^t (v - r) = 0 \tag{21.3}$$

Since we assume $v \neq 0$, the only possibility is if $v = r$. Equation (21.4) represents the general solution for Equation (21.1):

$$M_t = 3000 r^t \tag{21.4}$$

We know that if $r > 1$ we have exponential growth exponential decay when $r < 1$ exponential decay, consistent with our results above.

There is some comfort here: just like in continuous systems we find eigenvalues that determine the stability of the equilibrium solution. For discrete dynamical systems the stability is based on the value of an eigenvalue relative to 1 (not 0). Note: this is a good reminder to be aware if the model is based in continuous or discrete time!

21.3 Environmental stochasticity

It may be the case that environmental effects drastically change the net birth rate from one year to the next. For example during snowy winters the net birth rate changes because it is difficult to find food (Carroll 2013). For our purposes, let's say that in snowy winters r changes from 1.1 to 0.7. This would be a pretty drastic effect on the system - when $r = 1.1$ the moose population grows exponentially and when $r = 0.7$ the moose population decays exponentially.

A snowy winter occurs randomly. One way to model this randomness is to create a conditional statement based on the probability of it being snowy, defined on a scale from 0 to 1. How we implement this is by writing a function that draws a uniform random number each year and adjust the net birth rate:

```
# We use the snowfall_rate  as an input variable

moose_snow <- function(snowfall_prob) {
  out_moose <- array(M0, dim = N+1)
  for (i in 1:N) {
    r <- 1.1 # Normal net birth rate
    if (runif(1) < snowfall_prob) { # We are in a snowy winter
      r <- 0.7 # Decreased birth rate
    }
    out_moose[i + 1] <- r * out_moose[i]
  }
  return(out_moose)
}
```

Figure 21.2 displays different solution trajectories of the moose population over time for different probabilities of a deep snowpack.

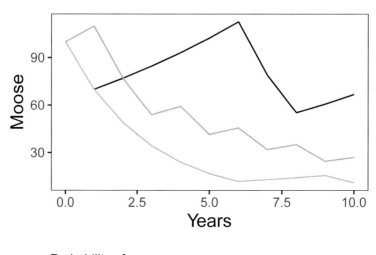

FIGURE 21.2 Moose populations with different probability of adjusting to deep snowpacks.

If you tried generating Figure 21.2 on your own you would not obtain the same figure. We are drawing random numbers for each year, so you should have different trajectories. While this may seem like a problem, one key thing that we will learn later in Chapter 22 is there is a stronger underlying signal when we compute *multiple* simulations and then compute an ensemble average.

As you can see when the probability of a snowy winter is very high ($p = 0.75$), the population decays exponentially. If that probability is lower, the moose population can still increase, but one bad year does knock the population down.

21.4   Discrete systems of equations

Another way to extend Equation (21.1) is to account for both adult (M) and juvenile (J) moose populations with Equation (21.5):

$$J_{t+1} = f \cdot M_t$$
$$M_{t+1} = g \cdot J_t + p \cdot M_t \tag{21.5}$$

Equation (21.5) is a little different from (21.1) because it includes juvenile and adult moose populations, which have the following parameters:

- f: represents the birth rate of new juvenile moose
- g: represents the maturation rate of juvenile moose
- p: represents the survival probability of adult moose

We can code up this model using R in the following way:

```r
M0 <- 900 # Initial population of adult moose
J0 <- 100 # Initial population of juvenile moose

N <- 10 # Number of years we run the simulation
moose_two_stage <- function(f, g, p) {

  # f: birth rate of new juvenile moose
  # g: maturation rate of juvenile moose
  # p: survival probability of adult moose

  # Create a data frame of moose to return
  out_moose <- tibble(
    years = 0:N,
    adult = M0,
    juvenile = J0
  )

  # And now the dynamics
  for (i in 1:N) {
```

```
    out_moose$juvenile[i + 1] <- f * out_moose$adult[i]
    out_moose$adult[i + 1] <-
      g * out_moose$juvenile[i] + p * out_moose$adult[i]
  }

  return(out_moose)
}
```

To simulate the dynamics we just call the function moose_two_stage and plot in Figure 21.3:

```
moose_two_stage_rates <- moose_two_stage(
  f = 0.5,
  g = 0.6,
  p = 0.7
)


ggplot(data = moose_two_stage_rates) +
  geom_line(aes(x = years, y = adult), color = "red") +
  geom_line(aes(x = years, y = juvenile), color = "blue") +
  labs(
    x = "Years",
    y = "Moose"
  )
```

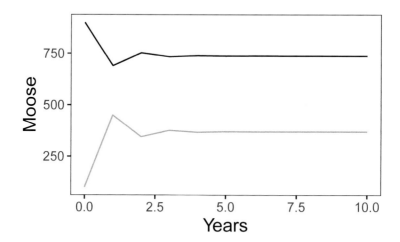

FIGURE 21.3 Simulation of a two stage moose population model.

Looking at Figure 21.3, it seems like both populations stabilize after a few years. We could further analyze this model for stable population states (in fact, it would be similar to determining eigenvalues as in Chapter 18). Additional extensions could also incorporate adjustments to the parameters f, g, and p in snow years (Exercise 21.5).

As you can see, introducing stochastic or random effects to a model yields some interesting (and perhaps more realistic) results. Next we will examine how computing can further explore stochastic models and how to generate expected patterns from all this randomness. Onward!

21.5   Exercises

Exercise 21.1. Re-run the moose population model with probabilities of adjusting to the deep snowpack at $p = 0$, 0.1, 0.9, and 1. How does adjusting the probability affect the moose population after 10 years?

Exercise 21.2. Modify the function `moose_snow` so that `runif(1) < snow-fall_prob` is changed to `runif(1) > snowfall_prob`. How does that code change the resulting solution trajectories in Figure 21.2? Why is this not the correct way to code changes in the net birth rate in deep snowpacks?

Exercise 21.3. Modify the two stage moose population model (Equation (21.5)) with the following parameters and plot the resulting adult and juvenile populations:

a. $f = 0.6$, $g = 0.6$, $p = 0.7$
b. $f = 0.5$, $g = 0.6$, $p = 0.4$
c. $f = 0.3$, $g = 0.6$, $p = 0.5$

You may assume $M_0 = 900$ and $J_0 = 100$.

Exercise 21.4. You are playing a casino game. If you win the game you earn $10. If you lose the game you lose your bet of $10. The probability of winning or losing is 50-50 (0.50). You decide to play the game 20 times and then cash out your net earnings.

a. Write code that is able to simulate this stochastic process. Plot your results.
b. Run this code five different times. What do you think your long term net earnings would be?
c. Now assume that you have a 40% chance of winning. Re-run your code to see how that affects your net earnings.

Exercise 21.5. Modify the two stage moose population model (Equation 21.5) to account for years with large snowdepths. In normal years, $f = 0.5$, $g = 0.6$, $p = 0.7$. However for snowy years, $f = 0.3$, $g = 0.6$, $p = 0.5$. Generate code that can account for these variable rates (similar to the moose population model). You may assume $M_0 = 900$, $J_0 = 100$, and N (the number of years) is 30. Plot simulations when the probability of snowy winters is $s = 0.05$ $s = 0.10$, or $s = 0.20$. Comment on the long-term dynamics of the moose for these simulations.

Exercise 21.6. A population grows according the the growth law $x_{t+1} = r_t x_t$.

a. Determine the general solution to this discrete dynamical system.
b. Plot a sample growth curve with $r_t = 0.86$ and $r_t = 1.16$, with $x_0 = 100$. Show your solution for $t = 50$ generations.
c. Now consider a model where $r_t = 0.86$ with probability $1/2$ and $r_t = 1.16$ with probability $1/2$. Write a function that will predict the population after $t = 50$. Show three or four different realizations of this stochastic process.

Exercise 21.7. (Inspired by Logan and Wolesensky (2009)) A rectangular preserve has area a. At one end of the boundary of the preserve (contained within the area), is a small band of land of area (a_b) from which animals disperse into the wilderness. Only animals at that eged disperse. Let u_t be the number of animals in a at any time t. The growth rate of all the animals in a is r. The rate at which animals disperse from the strip is proportional to the fraction of the animals in the edge band, with proportionality constant ϵ.

a. Draw a picture of the situation described above.
b. Explain why the equation that describes the dynamics is $u_{t+1} = r\, u_t - \epsilon \dfrac{a_b}{a} u_t$.
c. Determine conditions on the parameter r as a function of the other parameters under which the population is growing.

22

Simulating and Visualizing Randomness

In Chapter 21 we examined models for stochastic biological systems. These types of models are an introduction to the study of stochastic differential equations (SDEs). A common theme to SDEs is learning how to analyze and visualize randomness, broadly defined. In order to do that we will need to level up our skills to summarize a cohort of simulations over time. Let's get started!

22.1 Ensemble averages

Consider Figure 22.1, which shows the weather forecast for Kuopio, a city in Finland (Finnish Meteorological Institute 2021):[1]

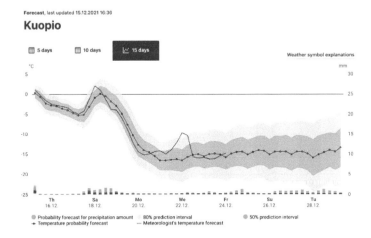

FIGURE 22.1 Long term weather forecast for Kuopio, a city in Finland, from the Finnish Weather Institute. Accessed 16-Dec 2021.

---

[1]While I could have picked any city, a lot of this textbook was written while I was on sabbatical in Kuopio. I highly recommend Finland as a country to visit.

TABLE 22.1 Simulations of a variable at different times.

t	sim1	sim2	sim3
1	3.39	2.16	1.93
2	0.16	3.99	4.53
3	2.92	4.85	3.15
4	4.05	3.68	4.65
5	1.65	0.23	3.82

Figure 22.1 shows a great example of what is called an *ensemble average*. The horizontal axis lists the time of day and the vertical axis is the temperature (the bar graph represents precipitation). The forecast temperature at a given point in time can have a range of outcomes, with the median of the distribution as the "temperature probability forecast"[2]. The red shading states that 80% of the outcomes fall in a given range, so while the median temperature on Monday, December 20 (labeled as Mo 20.12 - dates are represented as DAY-MONTH-YEAR) is $-10°C$, it may range between -16 and $-4°C$ (3 to 24 °F, brrr!). Based on the legends given, we would say the 80% confidence interval is between -10 to $-4°C$, or the models have 80% confidence for the temperature to be between that range of temperatures.

Because there may be different factors that alter the weather in a particular spot (e.g. the timing of a low pressure front, clouds, etc.) there are different possibilities for an outcome of the weather forecast. While it may seem like forecasting weather is impossible to do, sometimes these changes lead to small fluctuations in the forecasted weather at a given point. The ensemble average in Figure 22.1 becomes more uncertain (wider shading), as unforeseen events may drastically change the weather in the long term.

22.1.1   Spaghetti plots

Now let's focus on how to construct an ensemble average, but first let's start with a sample dataset. Consider the following data in Table 22.1. Notice how all the simulations (sim1, sim2, sim3) share the variable t in common, so it makes sense to plot them on the same axis in Figure 22.2. We call a plot of all of the simulations together a *spaghetti plot*, because, well, it can look like a bowl of spaghetti noodles was dumped all over the plotting space.

---

[2]Notice the meteorologist's temperature forecast on Wednesday, December 22 - sometimes what they predict may diverge from the model outcomes!

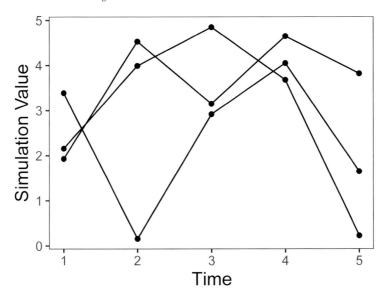

FIGURE 22.2 Spaghetti plot of the three simulations from Table 22.1.

While making the spaghetti plot isn't bad when you have three simulations, with (a lot) more simulations this would be a pain! An ensemble average computes *across* the rows in Table 22.1 (could be an average or a quantile) to generate a new column in the data. Building an ensemble average is a step by step process that involves a series of processes that transform a dataset where you will need to first pivot the dataset and then group and summarize.

22.1.2 Pivot

When you have multiple columns of a plot that you want to show together (such as a spaghetti plot) we can use a command called `pivot_longer` that gathers different columns together (also shown in Table 22.2).

```
my_table_long <- my_table %>%
  pivot_longer(cols = c("sim1":"sim3"),
               names_to = "name", values_to = "value")
```

Notice how the command `pivot_longer` takes the different simulations (`sim1`, `sim2`, `sim3`) and reassigns the column names to a new column called `name`, with values in the different columns appropriately assigned to the column `value`. This process called pivoting creates a new data frame, which makes it easier to generate the spaghetti plot (Figure 22.2). Try the following code out on your own to confirm this:

```
my_table_long %>%
  ggplot(aes(x = t, y = value, group = name)) +
```

```
n_sims <- 500 # The number of simulations

# Compute solutions
logistic_sim <- rerun(n_sims, c(x = runif(1, min = 0, max = 20))) %>%
  set_names(paste0("sim", 1:n_sims)) %>%
  map(~ euler(
    system_eq = logistic_eq,
    initial_condition = .x,
    parameters = params,
    deltaT = .05,
    n_steps = 200
  )) %>%
  map_dfr(~.x, .id = "simulation")

# Plot these all together
logistic_sim %>%
  ggplot(aes(x = t, y = x)) +
  geom_line(aes(color = simulation)) +
  ggtitle("Random initial conditions") +
  guides(color = "none")
```

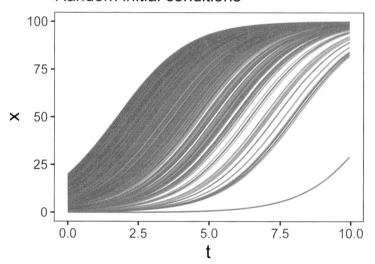

FIGURE 22.4 Spaghetti plot for logistic differential equation with 500 random initial conditions.

Wow! This spaghetti plot is really interesting - it should show how even though the initial conditions vary between $x = 0$ to $x = 20$, eventually all solutions flow to the carrying capacity $K = 100$ (which is a stable equilbrium solution). Initial conditions that start closer to $x = 0$ take longer, mainly because they are so close to the other equilibrium solution at $x = 0$ (which is an unstable equilibrium solution).

Ok, let's deconstruct this code line by line:

- `rerun(n_sims, c(x=runif(1,min=0,max=20)))` This line does two things: `x=runif(1,min=0,max=20)` makes a random initial condition, and the command `rerun` runs this again for `n_sims` times.
- `set_names(paste0("sim", 1:n_sims))` This line distinguishes between all the different simulations.
- `map(~ euler( ... )` You should be familiar with `euler`, but notice the pronoun `.x` that substitutes all the different initial conditions into Euler's method. The `map` function iterates over each of the simulations.
- `map_dfr(~ .x, .id = "simulation")` This line binds everything up together.

The resulting data frame should have three columns: - `simulation`: which one of the 500 simulations (`sim1`, `sim2`, etc ...) this corresponds to. - `t`: the value of the time - `x`: the output value of the variable x.

This code applies a new concept called *functional programming*. This is a powerful tool that allows you to perform the process of iteration (do the same thing repeatedly) with uncluttered code. We won't delve more into this here, but I encourage you to read about more functional programming concepts in Wickham and Grolemund (2017).

22.2.3 Summarize

Computing the ensemble average requires knowledge of how to use R to compute percentiles from a distribution of values. For our purposes here we will use the 95% confidence interval, so that means the 2.5 and 97.5 percentile (in which only 5% of the values will be outside of the specified interval), along with the median value (50th percentile). Let's take a look at the code for how to do that:

```
quantile_vals <- c(0.025, 0.5, 0.975)

logistic_quantile <- logistic_sim %>%
  group_by(t) %>%
  summarize(
    q_val = quantile(x,    # x is the column to compute the quantiles
      probs = quantile_vals
    ),
    q_name = quantile_vals
  ) %>%
```

Exercise 22.2. Read the Chapter 12 (tidy data) in Wickham and Grolemund (2017). In this chapter you will learn about tidy data. Explain what tidy data is and the potential uses for pivoting data wider or longer.

Exercise 22.3. Look at the documentation for quantile (remember you can type ?quantile at the command line to see the associated help for this function). Change the ensemble average in quantile_vals to compute the 25%, 50%, and 75% percentile for logistic_sim. Finally, produce a ensemble average plot of this percentile.

Exercise 22.4. Consider the logistic differential equation: $\dfrac{dx}{dt} = rx\left(1 - \dfrac{x}{K}\right)$. The function logistic_mod below takes the initial value problem $x(0) = 3$ and solves the differential equation.

a. Run logistic_mod(r=0.8,K=100) and plot its result.
b. Run 500 simulations with varying r chosen from a uniform distribution with minimum value of 0.4 and maximum value of 1.0. Create a spaghetti and ensemble average plot. Set $K = 100$.
c. Run 500 simulations with varying K chosen from a uniform distribution with minimum value of 50 and maximum value of 150. Create a spaghetti and ensemble average plot. Set $r = 0.8$.
d. Compare your results along with Figures 22.4 and 22.5. How does randomizing the initial condition or the parameters affect the results?

```
logistic_mod <- function(r,K) {
  logistic_eq <- c(dx ~ r * x * (1 - x / K)) # Define the rate equation

  params <- c(r=r,K=K) # Identify any parameters

  init_cond <- c(x = 3) # Initial condition
  soln <- euler(
    system_eq = logistic_eq,
    initial_condition = init_cond,
    parameters = params,
    deltaT = .05,
    n_steps = 200
  )

  return(soln)
}
```

Exercise 22.5. Using the data frame my_table, compare the following code below. The data frame table1 is skinny and long, and the second data frame table2 is called short and wide. Why did we need to make this data frame short and wide for plotting?

```
# First code chunk
table1 <- my_table %>%
  rowwise(t) %>%
  summarise(q_val = quantile(c_across(starts_with("sim")),
                                  probs = quantile_vals),
             q_name = quantile_vals)

# Second code chunk
table2 <- my_table %>%
  rowwise(t) %>%
  summarise(q_val = quantile(c_across(starts_with("sim")),
                                  probs = quantile_vals),
             q_name = quantile_vals) %>%
  pivot_wider(names_from = "q_name",values_from="q_val",
                  names_glue = "q{q_name}")
```

Exercise 22.6. Consider the following differential equation:

$$\frac{dx}{dt} = -y - x(x^2 + y^2 - 1)$$
$$\frac{dy}{dt} = x - y(x^2 + y^2 - 1)$$
(22.1)

a. Generate a phase plane for this differential equation. Store this phase plane in a variable called pp1. Set your x and y windows to be between -1 and 1.

b. The code below defines a function `limit_cycle_mod` that creates a solution trajectory of the differential equation. Super-impose a few different solution trajectories with random initial conditions onto your phase plane (pp1). Use initial conditions x0 and y0 between 0 and 1. Be sure to use the plot geom geom_path.

c. Modify the code from this chapter to run 50 different simulations with random initial conditions x0 and y0 between 0 and 1. *Note:* It may be helpful to include the code map(~ limit_cycle_mod(runif(1),runif(1))).

d. Plot the initial conditions from your simulation onto your phase plane. Isn't the result pretty?

```
limit_cycle_mod <- function(x0,y0) {
  limit_cycle_eq <- c(dx ~ -y-x*x*(x^2+y^2-1),
                  dy ~ x-y*(x^2+y^2-1) ) # Define the rate equation

  init_cond = c(x=x0,y=y0)
  soln <- rk4(
    system_eq = limit_cycle_eq,
    initial_condition = init_cond,
    deltaT = .05,
```

```
# Code for random variable r_q:
p <- runif(1)
if (p < 1 / 3) {
  x[i] <- x[i - 1] - 1
} else if (1 / 3 <= p & p < 2 / 3) {
  x[i] <- x[i - 1]
} else {
  x[i] <- x[i - 1] + 1
}
```

Exercise 23.9. In this exercise you will write code and simulate a two-dimensional random walk. In a given step you can either move (1) left, (2) right, (3) up, or (4) down. (You cannot move up and left for example). The random walk starts at $(x, y) = (0, 0)$. With $\Delta x = 1$, the random walk at step n can be described by $(x, y)^n = \sum_{s=1}^{n} r_d$, where r_d is one of the four motions, represented as a coordinate pair. (A movement up is $r_d = (0, 1)$ for example.)

a. Define a variable r_d that models the motion from step to step.
b. Modify the code for the one-dimensional random walk to incorporate this two-dimensional random walk. One way to do this is to create a variable y structured similar to x, and to have multiple `if` statements in the `for` loop that moves y.
c. Plot a few different realizations of your sample paths.
d. If we were to compute the mean and variance of the ensemble simulations, what do you think they would be?

24

Diffusion and Brownian Motion

Studying random walks in Chapter 23 led to some surprising results, namely that for an unbiased random walk the mean displacement was zero but the variance increased proportional to the step number. In this chapter we will revisit the random walk problem from another perspective that further strengthens its connection to understanding diffusion. Let's get started!

24.1 Random walk redux

The random walk derivation in Chapter 23 focused on the *position* of a particle on the random walk, based upon prescribed rules of moving to the left and the right. To revisit this random walk we consider the *probability* (between 0 and 1) that a particle is at position x in time t, denoted as $p(x, t)$. In other words, rather than focusing on where the particle *is*, we focus on the *chance* that the particle will be at a given spot.

A way to conceptualize a random walk is that any given position x, a particle can arrive to that position from either the left or the right (Figure 24.1):

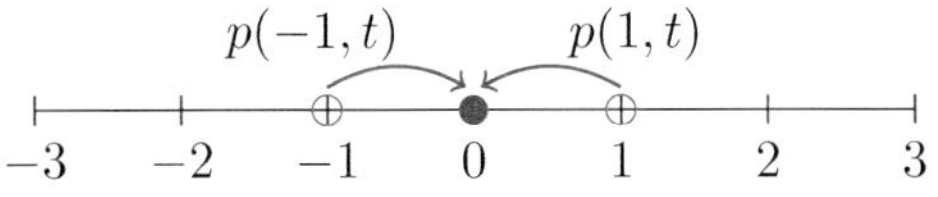

$$p(0, t + \Delta t) = \frac{1}{2}p(-1, t) + \frac{1}{2}p(1, t)$$

FIGURE 24.1 Schematic diagram for the one-dimensional random walk.

We can generalize Figure 24.1 further where the particle moves in increments Δx, as defined in Equation (24.1):

$$p(x, t + \Delta t) = \frac{1}{2}p(x - \Delta x, t) + \frac{1}{2}p(x + \Delta x, t) \qquad (24.1)$$

To analyze Equation (24.1) we apply Taylor approximations on each side of Equation (24.1). First let's do a locally linear approximation for $p(x, t + \Delta t)$:

$$p(x, t + \Delta t) \approx p(x, t) + \Delta t \cdot p_t, \tag{24.2}$$

where we have dropped the shorthand $p_t(x, t)$ as p_t. On the right hand side of Equation (24.1) we will compute the 2nd degree (quadratic) Taylor polynomial:

$$\frac{1}{2}p(x - \Delta x, t) \approx \frac{1}{2}p(x, t) - \frac{1}{2}\Delta x \cdot p_x + \frac{1}{4}(\Delta x)^2 \cdot p_{xx}$$

$$\frac{1}{2}p(x + \Delta x, t) \approx \frac{1}{2}p(x, t) + \frac{1}{2}\Delta x \cdot p_x + \frac{1}{4}(\Delta x)^2 \cdot p_{xx}$$

With these approximations we can re-write Equation (24.1) as Equation (24.3):

$$\Delta t \cdot p_t = \frac{1}{2}(\Delta x)^2 p_{xx} \rightarrow p_t = \frac{1}{2}\frac{(\Delta x)^2}{\Delta t} \cdot p_{xx} \tag{24.3}$$

Equation (24.3) is called a partial differential equation - what this means is that it is a differential equation with derivatives that depend on two variables (x and t (two derivatives). As studied in Chapter 23, Equation (24.3) is called the **diffusion equation**. In Equation (24.3) we can also define $D = \frac{1}{2}\frac{(\Delta x)^2}{\Delta t}$ so $p_t = D \cdot p_{xx}$.

The solution to Equation (24.3) is given by Equation (24.4).[1]

$$p(x, t) = \frac{1}{\sqrt{4\pi D t}}e^{-x^2/(4Dt)} \tag{24.4}$$

What Equation (24.4) represents is the probability that the particle is at the position x at time t. Figure 24.2 shows *profiles* for $p(x, t)$ when $D = 0.5$ at different values of t.

---

[1]Determining an exact solution to the diffusion equation requires more study in techniques of partial differential equations (see J. P. Keener (2021)).

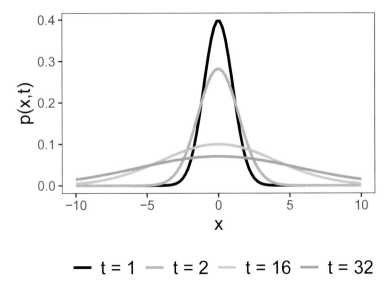

FIGURE 24.2 Profiles of $p(x,t)$ (Equation (24.4)) for different values of t with $D = 0.5$.

As you can see, as time increases the graph of $p(x,t)$ gets flatter - or more uniform. What this tells you is that the longer t increases it is less likely to find the particle at the origin.

24.1.1 Verifying the solution to the diffusion equation

Verifying that Equation (24.4) is the solution to Equation (24.3) is a good review of your multivariable calculus skills! As a first step to verifying this solution, let's take the partial derivative with respect to x and t.

First we will compute the partial derivative of p with respect to the variable x (represented as p_x):

$$p_x = \frac{\partial}{\partial x} \left(\frac{1}{\sqrt{4\pi Dt}} e^{-x^2/(4Dt)} \right)$$

$$= \frac{1}{\sqrt{4\pi Dt}} e^{-x^2/(4Dt)} \cdot \frac{-2x}{4Dt}$$

Notice something interesting here: $p_x = p(x,t) \cdot \left(\frac{-x}{2Dt} \right)$.

To compute the second derivative, we have the following expressions by applying the product rule:

$$p_{xx} = p_x \cdot \left(\frac{-x}{2Dt}\right) - p(x,t) \cdot \left(\frac{1}{2Dt}\right)$$

$$= p(x,t) \cdot \left(\frac{-x}{2Dt}\right) \cdot \left(\frac{-x}{2Dt}\right) - p(x,t) \cdot \left(\frac{1}{2Dt}\right)$$

$$= p(x,t) \left(\left(\frac{-x}{2Dt}\right)^2 - \left(\frac{1}{2Dt}\right)\right)$$

$$= p(x,t) \left(\frac{x^2 - 2Dt}{(2Dt)^2}\right).$$

So far so good. Now computing p_t gets a little tricky because this derivative involves both the product rule with the chain rule in two places (the variable t appears twice in the formula for $p(x,t)$). To aid in computing the derivative we identify two functions $f(t) = (4\pi Dt)^{-1/2}$ and $g(t) = -x^2 \cdot (4Dt)^{-1}$. This changes $p(x,t)$ into $p(x,t) = f(t) \cdot e^{g(t)}$. In this way $p_t = f'(t) \cdot e^{g(t)} + f(t) \cdot e^{g(t)} \cdot g'(t)$. Now we can focus on computing the individual derivatives $f'(t)$ and $g'(t)$ (after simplification - be sure to verify these on your own!):

$$f'(t) = -\frac{1}{2}(4\pi Dt)^{-3/2} \cdot 4\pi D = -2\pi D (4\pi Dt)^{-3/2}$$

$$g'(t) = x^2 (4Dt)^{-2} 4D = \frac{x^2}{4Dt^2}$$

Assembling these results together, we have the following:

$$p_t = f'(t) \cdot e^{g(t)} + f(t) \cdot e^{g(t)} \cdot g'(t)$$

$$= -2\pi D (4\pi Dt)^{-3/2} \cdot e^{-x^2/(4Dt)} + \frac{1}{\sqrt{4\pi Dt}} \cdot e^{-x^2/(4Dt)} \cdot \frac{x^2}{4Dt^2}$$

$$= \frac{1}{\sqrt{4\pi Dt}} \cdot e^{-x^2/(4Dt)} \left(-2\pi D (4\pi Dt)^{-1} + \frac{x^2}{4Dt^2}\right)$$

$$= \frac{1}{\sqrt{4\pi Dt}} \cdot e^{-x^2/(4Dt)} \left(-\frac{1}{2t} + \frac{x^2}{4Dt^2}\right)$$

$$= p(x,t) \left(-\frac{1}{2t} + \frac{x^2}{4Dt^2}\right)$$

Wow. Verifying that Equation (24.4) is a solution to the diffusion equation is getting complicated, but also notice that through algebraic simplification, $p_t = p(x,t) \left(\frac{x^2 - 2Dt}{4Dt^2}\right)$. When we compare p_t to Dp_{xx}, they are equal!

The connections between diffusion and probability are so strong. Equation (24.4) is related to the formula for a normal probability density function (Equation (9.1) from Chapter 9)! In this case, the standard deviation in

Equation (24.4) equals $\sqrt{2Dt}$ and is time dependent (see Exercise 24.2). Even though we approached the random walk differently here compared to Chapter 23, we also saw that the variance grew proportional to the time spent, so there is some consistency.

24.2   Simulating Brownian motion

Another name for the process of a particle undergoing small random movements is *Brownian Motion*. We can simulate Brownian motion similar to the random walk as discussed in Chapter 23. Brownian motion is connected to the diffusion equation (Equation (24.3)) and its solution (Equation (24.4)). These connections are helpful when simulating stochastic differential equations. To simulate Brownian motion we will also apply the workflow from Chapter 22 (Do once → Do several times → Summarize → Visualize).

24.2.1   Do once

First we define a function called `brownian_motion` that will compute a sample path given the:

- number of steps to run the stochastic process;
- diffusion coefficient D;
- timestep Δt;

a sample path will be computed (see Figure 24.3).

```
brownian_motion <- function(number_steps, D, deltaT) {
  # D: diffusion coefficient
  # deltaT: timestep length
  ### Set up vector of results
  x <- array(0, dim = number_steps)

  for (i in 2:number_steps) {
    x[i] <- x[i - 1] + sqrt(2 * D * deltaT) * rnorm(1)
  }

  out_x <- tibble(t = 0:(number_steps - 1), x)
  return(out_x)
}

# Run a sample trajectory and plot
try1 <- brownian_motion(100, 0.5, 0.1)
```

```
plot(try1, type = "l")
```

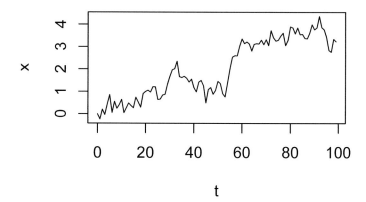

FIGURE 24.3 Sample trajectory for a realization of Brownian motion. The horizontal axis represents a step of the random walk.

24.2.2   Do several times

Once we have the function for Brownian motion defined we can then run this process several times and plot the spaghetti plot (try the following code out on your own):

```
number_steps <- 200 # Then number of steps in random walk
D <- 0.5 # The value of the diffusion coefficient
dt <- 0.1 # The timestep length

n_sims <- 500 # The number of simulations

# Compute solutions
brownian_motion_sim <- rerun(n_sims) %>%
  set_names(paste0("sim", 1:n_sims)) %>%
  map(~ brownian_motion(number_steps, D, dt)) %>%
  map_dfr(~.x, .id = "simulation")

# Plot these all together
ggplot(data = brownian_motion_sim, aes(x = t, y = x)) +
  geom_line(aes(color = simulation)) +
  ggtitle("Random Walk") +
  guides(color = "none")
```

24.2.3 Summarize and visualize

Finally, the 95% confidence interval is computed and plotted in Figure 24.4, using similar code from Chapter 22 to compute the ensemble average. Note that the horizontal axis is time so each step is scaled by dt.

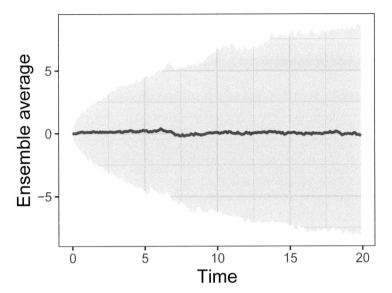

FIGURE 24.4 Ensemble average of 500 simulations for the random walk. Each step on the horizontal axis is scaled by dt.

I sure hope the results are very similar to ones generated in Chapter 23 (especially Figure 23.3) - this is no coincidence! With the ideas of a random walk developed here and in Chapter 23, we will now be able to understand and simulate how small changes in a variable or parameter affect the solutions to a differential equation. Looking ahead to Chapters 25 and 26, we will simulate stochastic processes using numerical methods (Euler's method in Chapter 4) with Brownian motion. Onward!

24.3 Exercises

Exercise 24.1. Through direct computation, verify the following calculations:

a. When $f(t) = \dfrac{1}{\sqrt{4\pi Dt}}$, then $f'(t) = -2\pi D(4\pi Dt)^{-3/2}$

b. When $g(t) = \dfrac{-x^2}{4Dt}$, then $g'(t) = \dfrac{x^2}{4Dt^2}$

c. Verify that $\left(-\dfrac{1}{2t} + \dfrac{x^2}{4Dt^2}\right) = \left(\dfrac{x^2 - 2Dt}{(2Dt)^2}\right)$

Exercise 24.2. The equation for the normal distribution is $f(x) = \dfrac{1}{\sqrt{2\pi}\sigma}e^{-(x-\mu)^2/(2\sigma^2)}$, with mean μ and variance σ^2. Examine the formula for the diffusion equation (Equation (24.4)) and compare it to the formula for the normal distribution. If Equation (24.4) represents a normal distribution, what do μ and σ^2 equal?

Exercise 24.3. For this problem you will investigate $p(x, t)$ (Equation (24.4)) with $D = \dfrac{1}{2}$.

a. Evaluate $\displaystyle\int_{-1}^{1} p(x, 10)\ dx$. Write a one sentence description of what this quantity represents.

b. Using desmos or some other numerical integrator, complete the following table:

Equation	Result
$\displaystyle\int_{-1}^{1} p(x, 10)\ dx =$	
$\displaystyle\int_{-1}^{1} p(x, 5)\ dx =$	
$\displaystyle\int_{-1}^{1} p(x, 2.5)\ dx =$	
$\displaystyle\int_{-1}^{1} p(x, 1)\ dx =$	
$\displaystyle\int_{-1}^{1} p(x, 0.1)\ dx =$	
$\displaystyle\int_{-1}^{1} p(x, 0.01)\ dx =$	
$\displaystyle\int_{-1}^{1} p(x, 0.001)\ dx =$	

c. Based on the evidence from your table, what would you say is the value of $\displaystyle\lim_{t\to 0^+}\int_{-1}^{1} p(x, t)\ dx$?

d. Now make graphs of $p(x, t)$ at each of the values of t in your table. What would you say is occuring in the graph as $\displaystyle\lim_{t\to 0^+} p(x, t)$? Does anything surprise you? (The results you computed here lead to the foundation of what is called the Dirac delta function.)

Exercise 24.4. Consider the function $p(x,t) = \dfrac{1}{\sqrt{4\pi Dt}} e^{-x^2/(4Dt)}$. Let $x = 1$.

a. Explain in your own words what the graph $p(1,t)$ represents as a function of t.

b. Graph several profiles of $p(1,t)$ when $D = 1$, 2, and 0.1. How does the value of D affect the profile?

Exercise 24.5. In statistics an approximation for the 95% confidence interval is twice the standard deviation. Confirm this by adding the curve $y = 2\sqrt{2Dt}$ to the ensemble average plot in Figure 24.4. Recall that D was equal to 0.5 and $\Delta t = 0.1$, so the horizontal axis will need to be scaled appropriately.

Exercise 24.6. Consider the function $p(x,t) = \dfrac{1}{\sqrt{\pi t}} e^{-x^2/t}$:

a. Using your differentiation skills compute the partial derivatives p_t, p_x, and p_{xx}.

b. Verify $p(x,t)$ is consistent with the diffusion equation $p_t = \dfrac{1}{4} p_{xx}$.

Exercise 24.7. Modify the code used to generate Figure 24.4 with $D = 10$, 1, 0.1, 0.01. Generally speaking, what happens to the resulting ensemble average when D is small or large? In which scenarios are stochastic effects more prevalent?

Exercise 24.8. For the one-dimensional random walk we discussed where there was an equal chance of moving to the left or the right. Here is a variation on this problem.

Let's assume there is a chance v that it moves to the left (position $x - \Delta x$), and therefore a chance is $1 - v$ that the particle remains at position x. The basic equation that describes the particle's position at position x and time $t + \Delta t$ is:

$$p(x, t + \Delta t) = (1 - v) \cdot p(x, t) + v \cdot p(x - \Delta x, t) \qquad (24.5)$$

Apply the techniques of local linearization in x and t to show that this random walk is used to derive the following partial differential equation, called the *advection equation*:

$$p_t = - \left(v \cdot \frac{\Delta x}{\Delta t} \right) \cdot p_x \qquad (24.6)$$

Note: you only need to expand this equation to first order

Exercise 24.9. Complete Exercise 23.9 if you haven't already. If Equation (24.4) (a normal distribution) is the solution to the one-dimensional diffusion equation, what do you think the solution would be in the bivariate case?

25

Simulating Stochastic Differential Equations

In this chapter we will begin to combine our knowledge of random walks to numerically simulate *stochastic differential equations*, or SDEs for short. Here is the good news: our previous work comes into focus. This chapter returns to a specific model you are familiar with (the logistic differential equation) and examines it stochastically. Hopefully this specific example will allow you to see how the methods developed here work in other contexts. Let's get started!

25.1 The stochastic logistic model

Equation (25.1) begins with the logistic differential equation, but written a little differently by multiplying the differential dt to the right hand side:

$$dx = rx \left(1 - \frac{x}{K}\right) dt \tag{25.1}$$

One way to interpret Equation (25.1) is that a small change is the variable x (denoted is dx), which is equal to the rate $rx \left(1 - \frac{x}{K}\right)$ multiplied by dt.

A direct way to incorporate stochastics is to modify Equation (25.1) by incorporating aspects of Brownian motion, as shown with Equation (25.2):

$$dx = \underbrace{rx \left(1 - \frac{x}{K}\right) dt}_{\text{Deterministic part}} + \underbrace{\sqrt{2D\,dt}\,\mathcal{N}(0, 1)}_{\text{Stochastic part}} \tag{25.2}$$

In Equation (25.2), D represents the diffusion coefficient and $\mathcal{N}(0, 1)$ signifies the normal distribution with mean zero and variance one.[1]

It may seem odd to express Equation (25.2) in this form (i.e. $dx = ...$ versus $\frac{dx}{dt} = ...$). However a good way to think of this stochastic differential equation is

---

[1] As a reminder, in R the command `rnorm(1)` draws a random number from the standard unit normal distribution.

that a small change in the variable x (represented by the term dx) is computed in two ways:

$$\text{Deterministic part: } rx\left(1 - \frac{x}{K}\right) dt$$
$$\text{Stochastic part: } \sqrt{2D\,dt}\,\mathcal{N}(0,1) \tag{25.3}$$

To simplify things somewhat we will represent $\sqrt{2D\,dt}\,\mathcal{N}(0,1)$ in Equation (25.2) with $dW(t)$, so that $dW(t) = \sqrt{2D\,dt}\,\mathcal{N}(0,1)$. The term $dW(t)$ can be thought of as similar to a stochastic differential equation $\dfrac{dW}{dt} = \sqrt{2D\,dt}$, with $W(0) = 0$. The solution $W(t)$ is also representative of a *Weiner process*. See Logan and Wolesensky (2009) For more information. In most cases a Weiner process does not include the term $\sqrt{2D}$ (effectively $D = \frac{1}{2}$). It is helpful to keep the term D as a control parameter for simulating the SDE (see Exercises 25.1 - 25.3).

How does the stochastic part of this differential equation change the solution trajectory? It turns out that the "exact" solutions to problems like these are difficult (we will study a sample of exact solutions to SDEs in Chapter 27). Rather than focus on exact solution techniques we will apply the workflow developed in Chapter 22 (Do once → Do several times → Summarize → Visualize) by simulating several solution trajectories and then taking the ensemble average at each of the time points.

25.2   The Euler-Maruyama method

One way to numerically solve a stochastic differential equation begins with a variation of Euler's method. The Euler-Maruyama method accounts for stochasticity and implements the random walk (Brownian motion). We will build this method up step by step.

Like Euler's method, the Euler-Maruyama method begins by writing the differential dx as a difference: $dx = x_{n+1} - x_n$, where n is the current step of the method. Likewise $dW(t) = W_{n+1} - W_n$, which represents one step of the random walk, but we approximate this difference by $\sqrt{2D\Delta t}\,\mathcal{N}(0,1)$, where Δt is the timestep length. Given Δt, diffusion coefficient D, and starting value x_0, we can define the following method.

- From the initial condition x_0, compute the value at the next time step (x_0), which for Equation (25.2) is:

$$x_1 = x_0 + rx_0\left(1 - \frac{x_0}{K}\right)\Delta t + \sqrt{2D\,\Delta t}\,\mathcal{N}(0,1)$$

- Repeat this iteration to step n, where $\mathcal{N}(0,1)$ is re-computed at each timestep:

$$x_n = x_{n-1} + r x_{n-1} \left(1 - \frac{x_{n-1}}{K}\right) \Delta t + \sqrt{2 D \Delta t}\,\mathcal{N}(0,1)$$

That is it! We can apply this numerical method for as many steps as we want. In the `demodelr` package the function `euler_stochastic` will apply the Euler-Maruyama method to a stochastic differential equation. Just like the functions `euler` or `rk4` there are some things that need to be set first:

- The size (Δt) of your timestep.
- The value of the diffusion coefficient D (we will discuss this later).
- The number of timesteps you wish to run the method. More timesteps means more computational time. If N is the number of timesteps, $\Delta t \cdot N$ is the total time.
- A function for our deterministic dynamics. For Equation (25.1) this equals $r x \left(1 - \frac{x}{K}\right)$.
- A function for our stochastic dynamics. For Equation (25.1) this equals 1.
- The values of the vector of parameters $\vec{\alpha}$. For the logistic differential equation we will take $r = 0.8$ and $K = 100$.

Sample code for this stochastic differential equation is shown below, with the resulting trajectory of the solution in Figure 25.1.

```
# Identify the deterministic and stochastic parts of the DE:
deterministic_logistic <- c(dx ~ r*x*(1-x/K))
stochastic_logistic <-  c(dx ~ 1)


# Identify the initial condition and any parameters
init_logistic <- c(x=3)
logistic_parameters <- c(r=0.8, K=100)   # parameters: a named vector


# Identify how long we run the simulation
deltaT_logistic <- .05     # timestep length
timesteps_logistic <- 200   # must be a number greater than 1


# Identify the standard deviation of the stochastic noise
D_logistic <- 1


# Do one simulation of this differential equation
logistic_out <- euler_stochastic(
  deterministic_rate = deterministic_logistic,
  stochastic_rate = stochastic_logistic,
  initial_condition = init_logistic,
  parameters = logistic_parameters,
  deltaT = deltaT_logistic,
```

```
    n_steps = timesteps_logistic,
    D = D_logistic
    )


# Plot out the solution
ggplot(data = logistic_out) +
    geom_line(aes(x=t,y=x))
```

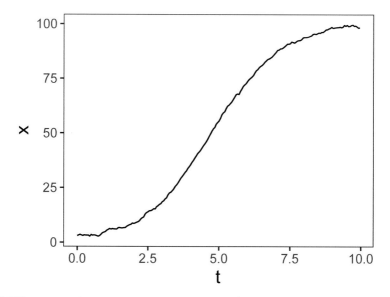

FIGURE 25.1 One realization of Equation (25.2).

Let's break the code down to generate Figure 25.1 step by step:

- We identify the deterministic and stochastic parts to our differential equation with the variables `deterministic_logistic` and `stochastic_logistic`. The same structure is used for Euler's method from Chapter 4.
- Similar to Euler's method we need to identify the initial conditions (`init_logistic`), parameters (`logistic_parameters`), Δt (`deltaT_logistic`), and number of timesteps (`timesteps_logistic`).
- The diffusion coefficient for the stochastic process (D) is represented with `D_logistic`.
- The command `euler_stochastic` does one realization of the Euler-Maruyama method.

25.2.1 Do several times

Figure 25.1 shows one sample trajectory of our solution, but there is benefit to running several simulations and then plotting out all the solution trajectories together. The code presented below accomplishes that task and makes a plot of all the solution trajectories (Figure 25.2). This code is similar to code presented in Chapter 22. As you may recall, the main engine of the code is contained in the `map( ~ euler_stochastic ... )` which re-runs the codes for the number of times specified in `n_sims`.

```r
# Many solutions
n_sims <- 100   # The number of simulations

# Compute solutions
logistic_run <- rerun(n_sims) %>%
  set_names(paste0("sim", 1:n_sims)) %>%
  map(~ euler_stochastic(deterministic_rate = deterministic_logistic,
                         stochastic_rate = stochastic_logistic,
                         initial_condition = init_logistic,
                         parameters = logistic_parameters,
                         deltaT = deltaT_logistic,
                         n_steps = timesteps_logistic,
                         D = D_log)) %>%
  map_dfr(~ .x, .id = "simulation")


# Plot these all together
ggplot(data = logistic_run) +
  geom_line(aes(x=t, y=x, color = simulation)) +
  ggtitle("Spaghetti plot for the logistic SDE") +
  guides(color="none")
```

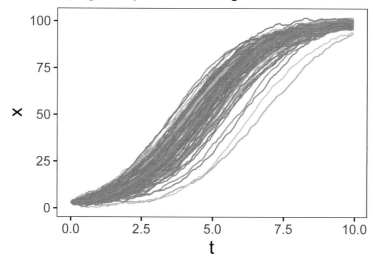

FIGURE 25.2 Several different realizations of Equation (25.2).

25.2.2   Summarize and Visualize

The code used to generate the ensemble average for the different simulations is shown below (try running this out on your own). This code is similar to ones presented in Chapter 22. The variable `summarized_logistic` first groups the simulations by the variable `t` in order to compute the quantiles across each of the simulations.

```
# Compute Quantiles and summarize
quantile_vals <- c(0.025, 0.5, 0.975)


### Summarize the variables
summarized_logistic <- logistic_run %>%
  group_by(t) %>%
  summarize(
    q_val = quantile(x,    # x is the column to compute the quantiles
                  probs = quantile_vals
    ),
    q_name = quantile_vals
  ) %>%
  pivot_wider(names_from = "q_name", values_from = "q_val",
            names_glue = "q{q_name}")

### Make the plot
ggplot(data = summarized_logistic) +
```

```
geom_line(aes(x = t, y = q0.5)) +
geom_ribbon(aes(x=t,ymin=q0.025,ymax=q0.975),alpha=0.2) +
ggtitle("Ensemble average plot for the logistic SDE")
```

25.3 Adding stochasticity to parameters

A second approach for modeling SDEs is to assume that the parameters are stochastic. For example, let's say that the growth rate r in the logistic differential equation is subject to stochastic effects. How we would implement this is by replacing r with $r +$ Noise :

$$dx = (r + \text{Noise}) \, x \left(1 - \frac{x}{K}\right) dt \qquad (25.4)$$

Now what we do is separate out the terms that are multiplied by "Noise" - they will form the stochastic part of the differential equation. The terms that aren't multipled by "Noise" form the deterministic part of the differential equation:

$$dx = rx \left(1 - \frac{x}{K}\right) dt + x \left(1 - \frac{x}{K}\right) \text{Noise} \, dt \qquad (25.5)$$

When we write Noise $dt = dW(t)$, then the deterministic and stochastic parts to Equation (25.5) are easily identified:

$$
\begin{aligned}
\text{Deterministic part: } & rx \left(1 - \frac{x}{K}\right) dt \\
\text{Stochastic part: } & x \left(1 - \frac{x}{K}\right) dW(t)
\end{aligned}
\qquad (25.6)
$$

There are a few things to notice with Equation (25.6). First, the deterministic part of the differential equation is what we would expect without incorporating the "Noise" term. Second, notice how the stochastic part of Equation (25.6) changed compared to Equation (25.3). Given these differences, let's see what happens when we simulate this SDE!

25.3.1 Do once

The following code will produce one realization of Equation (25.5), denoting the deterministic and stochastic parts as `deterministic_logistic_r` and `stochastic_logistic_r` respectively. We will also change the value of the diffusion coefficient D to equal 0.1. I encourage you to try this code on your own.

```
# Identify the deterministic and stochastic parts of the DE:
deterministic_logistic_r <- c(dx ~ r*x*(1-x/K))
stochastic_logistic_r <-  c(dx ~ x*(1-x/K))

# Identify the initial condition and any parameters
init_logistic <- c(x=3)
logistic_parameters <- c(K=100,r=0.8)    # parameters: a named vector

# Identify how long we run the simulation
deltaT_logistic <- .05      # timestep length
timesteps_logistic <- 200   # must be a number greater than 1

# Identify the standard deviation of the stochastic noise
D_logistic <- .1

# Do one simulation of this differential equation
logistic_out_r <- euler_stochastic(
  deterministic_rate = deterministic_logistic_r,
  stochastic_rate = stochastic_logistic_r,
  initial_condition = init_logistic,
  parameters = logistic_parameters,
  deltaT = deltaT_logistic,
  n_steps = timesteps_logistic,
  D = D_logistic
  )

# Plot out the solution
ggplot(data = logistic_out_r) +
  geom_line(aes(x=t,y=x))
```

25.3.2   Do several times

As we did before, we can run multiple iterations of Equation (25.5), which you can also try on your own. When you do this, does the resulting spaghetti plot look the same as or different from Figure 25.2? What could be some possible reasons for any differences?

```
# Many solutions
n_sims <- 100  # The number of simulations

# Compute solutions
logistic_run_r <- rerun(n_sims) %>%
  set_names(paste0("sim", 1:n_sims)) %>%
  map(~ euler_stochastic(deterministic_rate = deterministic_logistic_r,
                         stochastic_rate = stochastic_logistic_r,
```

```
                                 initial_condition = init_logistic,
                                 parameters = logistic_parameters,
                                 deltaT = deltaT_logistic,
                                 n_steps = timesteps_logistic,
                                 D = D_logistic)
) %>%
  map_dfr(~ .x, .id = "simulation")


# Plot these all together
  ggplot(data = logistic_run_r) +
  geom_line(aes(x=t, y=x, color = simulation)) +
  ggtitle("Spaghetti plot for the logistic SDE") +
  guides(color="none")
```

25.3.3   Summarize and Visualize

Now after running several different simulations we can plot the ensemble average, shown in Figure 25.3:

```
# Compute Quantiles and summarize
quantile_vals <- c(0.025, 0.5, 0.975)


# Summarize the variables
summarized_logistic <- logistic_run %>%
  group_by(t) %>%
  summarize(
    q_val = quantile(x,    # x is the column to compute the quantiles
                    probs = quantile_vals
    ),
    q_name = quantile_vals
  ) %>%
  pivot_wider(names_from = "q_name", values_from = "q_val",
              names_glue = "q{q_name}")


# Make the plot
ggplot(data = summarized_logistic) +
  geom_line(aes(x = t, y = q0.5)) +
  geom_ribbon(aes(x=t,ymin=q0.025,ymax=q0.975),alpha=0.2) +
  ggtitle("Ensemble average plot for the logistic SDE")
```

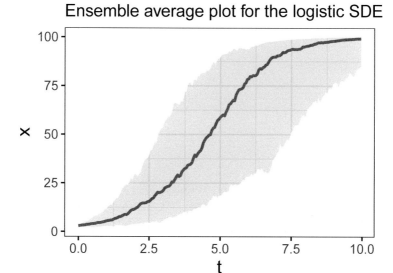

FIGURE 25.3 Ensemble average plot for Equation (25.5).

The results you obtain in Figure 25.3 might look similar to the ensemble average from simulations of Equation (25.2), but perhaps with more variability (represented in the shading for the 95% confidence interval). Letting r be a stochastic parameter affects how quickly the solution x increases to the carrying capacity at $x = 100$.

25.3.4   When you may have odd results

Let's discuss failure and other oddities when simulating SDEs. Figure 25.4 shows one odd realization of Equation (25.5):

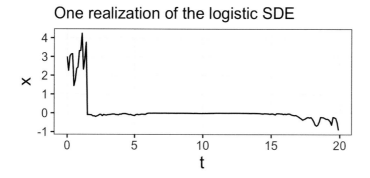

FIGURE 25.4 One odd realization of Equation 25.5.

Notice in Figure 25.4 how the variable x has values that are negative. Is this a problem? ... Maybe. As with analyzing (non-stochastic) differential equations, unanticipated results may be a signal of the following:

- You may have an error in the coding of the SDE (always double-check your work!)
- You chose too large of a timestep Δt. For Brownian motion, the variance grows proportional to Δt, so larger timesteps mean more variability when you simulate a random variable, which leads to larger stochastic jumps. (To be fair, large values of Δt are also a shortcoming of Euler's method.)
- Similar to the last point, the value of D may be too large. Recall that D is the diffusion coefficient and affects the rate of spread. Try setting $D = 0$ and seeing if that produces a result in line with your expectation from the (non-stochastic) differential equation.
- The Euler-Maruyama may cause the variable to move across an equilibrium solution, thereby undergoing a change in the long-term behavior.[2] For the logistic equation, we know that $x = 0$ is a unstable equilibrium solution. If for this instance x becomes negative in Figure 25.4, it will move (quickly) away from $x = 0$ in the negative direction.
- Sometimes you may get NaN values in your simulations. You may still be able to compute the ensemble average, but you will need add to the following `na.rm = TRUE` in your `quantile` command (using the variable `logistic_run` as an example):

```
# Summarize the variables
summarized_logistic <- logistic_run %>%
  group_by(t) %>%
  summarize(
    q_val = quantile(x,    # x is the column to compute the quantiles
               probs = quantile_vals
    na.rm = TRUE),   # Note the include of na.rm = TRUE
    q_name = quantile_vals
  ) %>%
  pivot_wider(names_from = "q_name", values_from = "q_val",
              names_glue = "q{q_name}")
```

The last point is not a fault of the numerical method, but rather a feature of the differential equation. It is *always* helpful to understand your underlying dynamics before you start to implement a stochastic process!

---

[2]Dynamically, a variable cannot cross an equilibrium solution. If a variable x reaches an equilibrium solution at a finite time t, then it would remain at that equilibrium solution (see Exercise 5.10).

25.4   Systems of stochastic differential equations

To incorporate stochastic effects with a system of differential equations the process is similar to above, with a few changes. First, we need to keep track of the deterministic and stochastic parts *for each variable*. Second, when we summarize our results in computing the ensemble averages we also need to group by each of the variables. Let's take a look at an example.

Let's revisit the tourism model from Chapter 13 (Sinay and Sinay 2006). This model described in Equation (25.7) relies on two non-dimensonal scaled variables: R, which is the amount of the resource (as a percentage), and V, the percentage of visitors that could visit (also as a percentage).

$$\frac{dR}{dt} = R \cdot (1 - R) - aV$$
$$\frac{dV}{dt} = b \cdot V \cdot (R - V) \tag{25.7}$$

Equation (25.7) has two parameters a and b, which relate to how the resource is used up as visitors come (a) and how as the visitors increase, word of mouth leads to a negative effect of it being too crowded (b). Sinay and Sinay (2006) reported $a = 0.15$ and $b = 0.3316$. Let's assume that b is stochastic, leading to the following deterministic and stochastic parts to Equation (25.7):

- Deterministic part for $\dfrac{dR}{dt}$: $R \cdot (1 - R) - aV$
- Stochastic part for $\dfrac{dR}{dt}$: 0
- Deterministic part for $\dfrac{dV}{dt}$: $b \cdot V \cdot (R - V)$
- Stochastic part for $\dfrac{dV}{dt}$: $V \cdot (R - V)$

Now we will apply the established workflow. The code to generate one realization of this stochastic process (the "Do once" step) is shown below (try this out on your own):

```
# Identify the deterministic and stochastic parts of the DE:
deterministic_tourism<- c(dr ~ R*(1-R)-a*V, dv ~ b*V*(R-V))
stochastic_tourism <-  c(dr ~ 0, dv ~ V*(R-V))

# Identify the initial condition and any parameters
init_tourism <- c(R = 0.995, V = 0.00167)
tourism_parameters <- c(a = 0.15, b = 0.3316)    #

deltaT_tourism <- .5 # timestep length
```

```
timeSteps_tourism <- 200 # must be a number greater than 1

# Identify the diffusion coefficient
D_tourism <- .05

# Do one simulation of this differential equation
tourism_out <- euler_stochastic(
  deterministic_rate = deterministic_tourism,
  stochastic_rate = stochastic_tourism,
  initial_condition = init_tourism,
  parameters = tourism_parameters,
  deltaT = deltaT_tourism,
  n_steps = timeSteps_tourism,
  D = D_tourism
  )

# We will pivot the data to ease in plotting:
tourism_revised <- tourism_out %>%
  pivot_longer(cols=c("R","V"))

# Plot out the solution
ggplot(data = tourism_revised) +
  geom_line(aes(x=t,y=value)) +
  facet_grid(.~name)
```

Note the additional creation of the variable `tourism_revised` and the command `facet_grid` in the plotting. Let's break this down:

- The data frame `tourism_out` has three variables: t, R, and V. Because we have more than one variable, we need to do an additional step in pivoting longer. (`pivot_longer(cols=c("r","v"))`) The data frame `tourism_revised` has more rows, but the second column (called `name`) contains the name of each variable), and the third column (`value`) has the associated value of each variable at a given point in time.
- In order to plot these variables in a multipanel plot we use `facet_grid`. Think of `facet_grid(.~name)` as a row by column display. We want the columns to be the instances of the variable `name`, with one row (`.`).

The "Do several times" step follows a similar process as previous examples. The code to generate and plot multiple realizations of this stochastic process is below and also displayed with Figure 25.5.

```
# Many solutions
n_sims <- 100   # The number of simulations

# Compute solutions
```

```
tourism_run <- rerun(n_sims) %>%
  set_names(paste0("sim", 1:n_sims)) %>%
  map(~ euler_stochastic(
    deterministic_rate = deterministic_tourism,
    stochastic_rate = stochastic_tourism,
    initial_condition = init_tourism,
    parameters = tourism_parameters,
    deltaT = deltaT_tourism,
    n_steps = timeSteps_tourism,
    D = D_tourism)
    ) %>%
  map_dfr(~ .x, .id = "simulation")

# We will pivot the data to ease in plotting and computing:
tourism_run_revised <- tourism_run %>%
  pivot_longer(cols=c("R","V"))

# Plot these all together
ggplot(data = tourism_run_revised) +
  geom_line(aes(x=t, y=value, color = simulation)) +
  facet_grid(.~name) +
  guides(color="none")
```

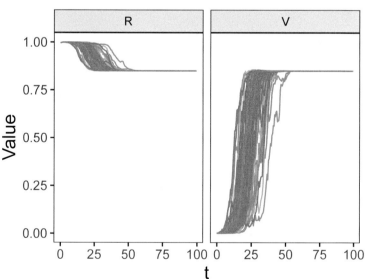

FIGURE 25.5 Spaghetti plot for many realizations of Equation (25.7).

Notice how the different realizations in Figure 25.5 show eventually the variables R and V approaching a steady-state value, but when b is stochastic it affects how quickly the steady state is approached. The variability in R and V (especially during the intervals $0 \leq t \leq 50$) may be important for how quickly this resource gets utilized.

The final steps (Summarize and Visualize) can be down together and as you may have suspected, have a similar process to the previous examples. However when computing the ensemble average plot, there are three changes because we pivoted our results in the variable `tourism_run_revised`:

- When we group by our variables to compute the ensemble average we use `group_by(t,name)` so that we organize by each time `t` and the variables in our system (gathered in the column called `name`).
- When computing the ensemble average, we will need to specify that we are computing the quantile of the variable `value` in our pivoted data frame (`tourism_run_revised`).
- Finally, as in Figure 25.5, we need to apply the `facet_grid` command.

The code to generate this ensemble average plot is shown below. Try this code out on your own to generate an ensemble average plot.

```
# Compute Quantiles and summarize
quantile_vals <- c(0.025, 0.5, 0.975)

# Summarize the variables
summarized_tourism <- tourism_run_revised %>%
  group_by(t,name) %>%  # We also include grouping by each variable.
  summarize(
    q_val = quantile(value, # the column to compute quantiles
                     probs = quantile_vals,
                     na.rm = TRUE  # Remove NA vlaues
    ),
    q_name = quantile_vals
  ) %>%
  pivot_wider(names_from = "q_name", values_from = "q_val",
              names_glue = "q{q_name}")

# Make the plot
ggplot(data = summarized_tourism) +
  geom_line(aes(x = t, y = q0.5)) +
  geom_ribbon(aes(x = t, ymin = q0.025, ymax = q0.975),alpha = 0.2) +
  facet_grid(.~name) +
  ggtitle("Ensemble average plot")
```

$$\frac{dI}{dt} = b(N - I)I - rI \tag{25.8}$$

a. Determine the equilibrium solutions for this model. As a bonus, analyze the stability of the equilibrium solutions. You will need to assume that all parameters are positive and $bN - r > 0$.

b. Assuming $N = 1000$, $r = 0.01$, and $b = 0.005$, $I(0) = 1$, apply Euler's method to simulate this differential equation over two weeks with $\Delta t = 0.1$ days. Show the plot of your result.

c. Assume the transmission rate b is stochastic. Write down this stochastic differential equation. Do 500 simulations of this stochastic process with with $D = 1 \cdot 10^{-6}$. Generate a spaghetti plot of your results. Contrast this result to the deterministic solution.

d. Assume the recovery rate r is stochastic. Write down this stochastic differential equation. Do 500 simulations of this stochastic process with $D = 1 \cdot 10^{-3}$. Generate a spaghetti plot of your results. Contrast this result to the deterministic solution.

Exercise 25.8. Organisms that live in a saline environment biochemically maintain the amount of salt in their blood stream. An equation that represents the level of S (as a percent) in the blood is the following:

$$\frac{dS}{dt} = I + p \cdot (W - S),$$

where the parameter I represents the active uptake of salt (% / hour), p is the permeability of the skin (hour^{-1}), and W is the salinity in the water (as a percent). Use this information to answer the following questions:

a. When $S(0) = S_0$, apply techniques from Chapter 7 to determine an exact solution for this initial value problem.

b. Set $I = 0.1$, $p = 0.05$, $W = 0.4$, $S_0 = 0.6$. Make a plot of your solution for $0 \le t \le 20$.

c. What is a corresponding stochastic differential equation when the parameter p is stochastic?

d. Set $I = 0.1$, $p = 0.05$, $W = 0.4$, $S_0 = 0.6$. With $\Delta t = 0.05$ and for 400 timesteps, with $D = 0.1$, simulate this stochastic process. With an ensemble of 200 realizations compare the generated ensemble average to your deterministic solution.

Exercise 25.9. (Inspired by Munz et al. (2009)) Consider the following model for zombie population dynamics:

$$\frac{dS}{dt} = -\beta SZ - \delta S$$

$$\frac{dZ}{dt} = \beta SZ + \xi R - \alpha SZ \qquad (25.9)$$

$$\frac{dR}{dt} = \delta S + \alpha SZ - \xi R$$

a. Let's assume the transmission rate β is a stochastic parameter. With this assumption, group each differential equation into two parts: terms not involving noise (the deterministic part) and terms that are multiplied by noise (the stochastic part)

- Deterministic part for $\dfrac{dS}{dt}$:
- Stochastic part for $\dfrac{dS}{dt}$:
- Deterministic part for $\dfrac{dZ}{dt}$:
- Stochastic part for $\dfrac{dZ}{dt}$:
- Deterministic part for $\dfrac{dR}{dt}$:
- Stochastic part for $\dfrac{dR}{dt}$:

b. Apply the Euler-Maruyama method to generate an ensemble avearage plot with the following values:

- $D = 5 \cdot 10^{-6}$
- $\Delta t = 0.05$.
- Timesteps: 1000.
- $\beta = 0.0095$, $\delta = 0.0001$,$\xi = 0.1$, $\alpha = 0.005$.
- Initial condition: $S(0) = 499$, $Z(0) = 1$, $R(0) = 0$.
- Set the number of simulations to be 100.

c. How does making β stochastic affect the disease transmission?

Exercise 25.10. Evaluate results from stochastic simulation of Equation (25.7) when the parameter a is stochastic. You will need to determine an appropriate value of D with the values of the parameters and timestep given. Contrast your findings with the results presented in Figure 25.5.

```
  x <- array(x0, dim = number_steps)

  # Iterate through this random process.
  for (i in 2:number_steps) {
    x[i] <- x[i - 1] + rnorm(1) * sqrt(dt)
  }

  # Create the time vector
  t <- seq(0, length.out = number_steps, by = dt)
  out_x <- tibble(t, x)
  return(out_x)
}

out <- sde(1000, .2)

ggplot(data = out) +
  geom_line(aes(x = t, y = x))
```

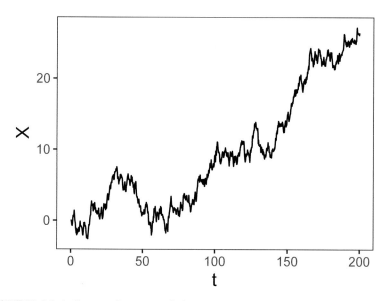

FIGURE 26.1 One realization of the stochastic differential equation $dX = 2\,dW(t)$.

Let's highlight two parts from the function `sde`:

1. Notice the line `x[i] <- x[i-1] + rnorm(1)*sqrt(dt)` where we iterate through the random process with the term $\sqrt{dt}$, similar to what we did in Chapter 25.

2. To produce the correct time intervals in Figure 26.1 we defined a time vector: t <- seq(0,length.out=number_steps,by=dt).

When we repeat this process several times and plot of the results, we have the following ensemble average in Figure 26.2:

```r
# Many solutions
n_sims <- 1000 # The number of simulations

# Compute solutions
sde_run <- rerun(n_sims) %>%
  set_names(paste0("sim", 1:n_sims)) %>%
  map(~ sde(1000, .2)) %>%
  map_dfr(~.x, .id = "simulation")


# Compute Quantiles and summarize
quantile_vals <- c(0.025, 0.5, 0.975)


# Summarize the variables
summarized_sde <- sde_run %>%
  group_by(t) %>%
  summarize(
    q_val = quantile(x, # x is the column to compute the quantiles
      probs = quantile_vals,
      na.rm = TRUE # remove NA values
    ),
    q_name = quantile_vals
  ) %>%
  pivot_wider(
    names_from = "q_name", values_from = "q_val",
    names_glue = "q{q_name}"
  )

# Make the plot
ggplot(data = summarized_sde) +
  geom_line(aes(x = t, y = q0.5)) +
  geom_ribbon(aes(x = t, ymin = q0.025, ymax = q0.975), alpha = 0.2)
```

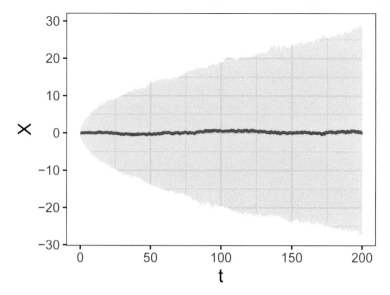

FIGURE 26.2 Ensemble average of 1000 realizations of the stochastic differential equation $dX = dW(t)$.

I sure hope (again!) the results are very similar to ones generated in Chapters 23 and Chapter 24 (especially Figures 23.3 and 24.4) - this is no coincidence! Figure 24.4 is another simulation of Brownian motion with $D = 1/2$.

26.1.2   Stochastics with drift

The second example is a modification where $a(X,t) = 2$ and $b(X,t) = 1$ in Equation (26.1), yielding $dX = 0.2\,dt + dW(t)$. To simulate this stochastic process you can easily modify this approach by modifying the function sde, which I will call sde_v2. The code to simulate the stochastic process and visualize the ensemble average (Figure 26.3) is shown below:

```
sde_v2 <- function(number_steps, dt) {
  a <- 2
  b <- 1 # The value of b
  x0 <- 0 # The initial condition


  ### Set up vector of results
  x <- array(x0, dim = number_steps)

  for (i in 2:number_steps) {
    x[i] <- x[i - 1] + a * dt + b * rnorm(1) * sqrt(dt)
  }
```

```r
  # Create the time vectror
  t <- seq(0, length.out = number_steps, by = dt)
  out_x <- tibble(t, x)
  return(out_x)
}


# Many solutions
n_sims <- 1000 # The number of simulations

# Compute solutions
sde_v2_run <- rerun(n_sims) %>%
  set_names(paste0("sim", 1:n_sims)) %>%
  map(~ sde_v2(1000, .2)) %>%
  map_dfr(~.x, .id = "simulation")


# Compute Quantiles and summarize
quantile_vals <- c(0.025, 0.5, 0.975)


### Summarize the variables
summarized_sde_v2 <- sde_v2_run %>%
  group_by(t) %>%
  summarize(
    q_val = quantile(x, # x is the column to compute the quantiles
      probs = quantile_vals,
      na.rm = TRUE # Remove NA values
    ),
    q_name = quantile_vals
  ) %>%
  pivot_wider(
    names_from = "q_name", values_from = "q_val",
    names_glue = "q{q_name}"
  )

### Make the plot
ggplot(data = summarized_sde_v2) +
  geom_line(aes(x = t, y = q0.5)) +
  geom_ribbon(aes(x = t, ymin = q0.025, ymax = q0.975), alpha = 0.2)
```

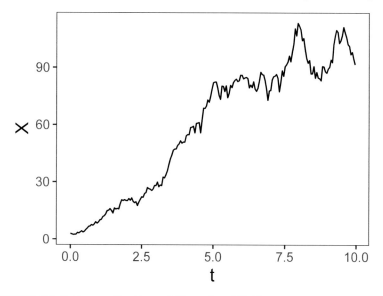

FIGURE 26.5 One realization of Equation (26.6).

The following code will produce a spaghetti plot from 100 different simulations. Try this code out on your own.

```
# Many solutions
n_sims <- 100 # The number of simulations

# Compute solutions
logistic_sim_r <- rerun(n_sims) %>%
  set_names(paste0("sim", 1:n_sims)) %>%
  map(~ euler_stochastic(
    deterministic_rate = deterministic_rate_log,
    stochastic_rate = stochastic_rate_log,
    initial_condition = init_log,
    parameters = parameters_log,
    deltaT = deltaT_log,
    n_steps = time_steps_log,
    D = D_log
  )) %>%
  map_dfr(~.x, .id = "simulation")


# Plot these all together
ggplot(data = logistic_sim_r) +
  geom_line(aes(x = t, y = x, color = simulation)) +
  guides(color = "none")
```

Finally, Figure 26.6 displays the ensemble average plot from all the simulations. Figure 26.6 also includes the solution to the the logistic differential equation for comparison.

```
# Compute Quantiles and summarize
quantile_vals <- c(0.025, 0.5, 0.975)


# Summarize the variables
summarized_logistic <- logistic_sim_r %>%
  group_by(t) %>%
  summarize(
    q_val = quantile(x, # x is the column to compute the quantiles
      probs = quantile_vals,
      na.rm = TRUE
    ),
    q_name = quantile_vals
  ) %>%
  pivot_wider(
    names_from = "q_name", values_from = "q_val",
    names_glue = "q{q_name}"
  )

logistic_solution <- tibble(
  t = seq(0, 10, length.out = 100),
  x = 100 / (1 + 97 / 3 * exp(-0.8 * t))
)


# Make the plot
ggplot(data = summarized_logistic) +
  geom_line(aes(x = t, y = q0.5),
            color = "red", size = 1) +
  geom_ribbon(aes(x = t, ymin = q0.025, ymax = q0.975),
              alpha = 0.2, fill = "red") +
  geom_line(data = logistic_solution, aes(x = t, y = x),
            size = 1, linetype = "dashed")
```

Generate a spaghetti plot and ensemble average of your simluation results.

Exercise 26.2. Return to the simulation of the logistic differential equation in this chapter. To generate Figure 26.6 we set $D = 1$. What happens to the resulting ensemble average plots when $D = 0$, 0.01, 0.1, 1, 10? You may use the following values:

- Set the number of simulations to be 100.
- Initial condition: $x(0) = 3$
- Parameters: $r = 0.8$, $K = 100$.
- $\Delta t = 0.05$ for 200 time steps.

Exercise 26.3. For the logistic differential equation consider the following splitting of $\alpha(x)$ and $\delta(x)$ as a birth-death process:

$$\alpha(x) = rx - \frac{rx^2}{2K}$$
$$\delta(x) = \frac{rx^2}{2K} \tag{26.8}$$

Simulate this SDE with the following values:

- Initial condition: $x(0) = 3$
- Parameters: $r = 0.8$, $K = 100$.
- $\Delta t = 0.05$ for 200 time steps.
- $D = 1$.

Generate an ensemble average plot. How does this SDE compare to Figure 26.6?

Exercise 26.4. Let $S(t)$ denote the cumulative snowfall at a location at time t, which we will assume to be a random process. Assume that probability of the change in the cumulative amount of snow from day t to day $t + \Delta t$ is the following:

change	probability
$\Delta S = \sigma$	$\lambda \Delta t$
$\Delta S = 0$	$1 - \lambda \Delta t$

The parameter λ represents the frequency of snowfall and σ the amount of the snowfall in inches. For example, during January in Minneapolis, Minnesota, the probability λ of it snowing 4 inches or more is 0.016, with $\sigma = 4$. (This assumes a Poisson process with rate $= 0.5/31$, according to the Minnesota

DNR.[1]). The stochastic differential equation generated by this process is
$dS = \lambda\sigma\, dt + \sqrt{\lambda\sigma^2}\, dW(t) = .064\, dt + .506\, dW(t)\$$.

a. With this information, what is $E[\Delta S]$ and the variance of ΔS?
b. Simulate and summarize this stochastic process. Use $S(0) = 0$ and run 500 simulations of this stochastic process. Simulate this process for a month, using $\Delta t = 0.1$ for 300 timesteps and with $D = 1$. Show the resulting spaghetti plot and interpret your results.

Exercise 26.5. Consider the stochastic differential equation $dS = (1 - S)\, dt + \sigma\, dW(t)$, where σ controls the amount of stochastic noise. For this stochastic differential equation what is $E[S]$ and $\mathrm{Var}(S)$?

Exercise 26.6. Consider the equation

$$\Delta x = \alpha(x)\, \Delta t - \delta(x)\, \Delta t$$

If we consider Δx to be a random variable, show that the expected value μ equals $\alpha(x)\, \Delta t - \delta(x)\, \Delta t$ and the variance σ^2, to first order, equals $\alpha(x)\, \Delta t + \delta(x)\, \Delta t$.

---

[1] https://www.dnr.state.mn.us/climate/twin_cities/snow_event_counts.html

discussion and derivation of the Fokker-Planck equation). However, let's build up understanding of Equation (27.2) through some examples.

27.1.1   Diffusion (again)

Consider the SDE $dx = dW(t)$ with $x(0) = 0$ and apply the Fokker-Planck equation to characterize the solution $p(x, t)$. We know from Chapter 24 that SDE $dx = dW(t)$ characterizes Brownian motion. When we compare this SDE to the Fokker-Planck equation and Equation (27.1), we have $a(x, t) = 0$ and $b(x, t) = 1$, yielding Equation (27.1.1):

$$p_t = \frac{1}{2}p_{xx}.$$

This equation should look familiar - it is the partial differential equation for diffusion (Equation (24.3))![1] The solution to this SDE is given by Equation (27.3). Figure 27.1 shows the evolution of $p(x, t)$ in time.

$$p(x, t) = \frac{1}{\sqrt{2\pi t}} e^{-x^2/(2t)} \tag{27.3}$$

One way to describe Equation (27.3) is a normally distributed random variable, with $E[p(x, t)] = 0$ and σ^2 (the variance) equal to t. Notice how the mean and variance for Equation (27.3) connect back to our previous work with random walks and diffusion in Chapters 23 and 24. Namely, simulations and random walk mathematics showed that the expected value of a random walk or Brownian motion was zero and the variance grew in time, which is the same for this SDE.

---

[1]To remind you, the solution to $p_t = Dp_{xx}$ is $p(x, t) = \dfrac{1}{\sqrt{4\pi Dt}} e^{-x^2/(4Dt)}$. So in this case $D = 1/2$.

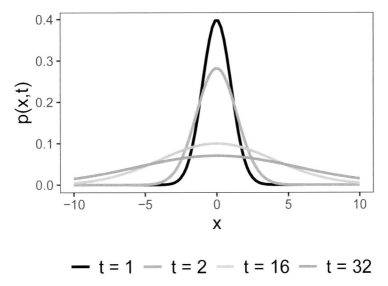

FIGURE 27.1 Profiles for the solution to SDE $dx = dW(t)$ (given by Equation (27.3)) for different values of t.

Let's discuss the initial condition for Equation (27.3). Our SDE had the initial condition $x(0) = 0$. How this initial condition translates to Equation (27.3) is that $x(0)$ is the same as $p(x, 0)$. For this example the initial condition $p(x, 0)$ is a special function called the Dirac delta function[2], written as $p(x, 0) = \delta(x)$. The function $\delta(x)$ is a special type of probability density function, which you may study in a course that explores the theory of functions. Applications of the Dirac delta function include modeling an impulse (such as the release of a particle from a specific point) and watching its evolution in time.

27.1.2 Diffusion with drift

Now that we have a handle on the SDE $dx = dW(t)$, let's extend this next example a little more. Consider the SDE $dx = r\,dt + \sigma\,dW(t)$, where r and σ are constants. As a first step, let's examine the deterministic equation: $dx = r\,dt$, which has a linear function $x(t) = rt + x_0$ as its solution.[3] As before, the initial condition for this SDE is $p(x, 0) = \delta(x)$.

Applying Equation (27.2) to this SDE we obtain Equation (27.4):

$$p_t = -r\,p_x + \frac{\sigma^2}{2}\,p_{xx} \tag{27.4}$$

---

[2]https://en.wikipedia.org/wiki/Dirac_delta_function

[3]This example is similar to the SDE $dX = 0.2\,dt + dW(t)$ in Chapter 26, where $r = 0.2$ and $\sigma = 1$. In that example we found that the variance σ^2 grew linearly in time.

References

Akaike, Hirotugu. 1974. "A New Look at the Statistical Model Identification." *IEEE Transactions on Automatic Control* 19 (6): 716–23. `https://doi.or g/10.1109/TAC.1974.1100705`.

Berg, Howard. 1993. *Random Walks in Biology*. Princeton, New Jersey: Princeton University Press.

Berg, Hugo van den. 2011. *Mathematical Models of Biological Systems*. Oxford, New York: Oxford University Press.

Beven, Keith, and Jim Freer. 2001. "Equifinality, Data Assimilation, and Uncertainty Estimation in Mechanistic Modelling of Complex Environmental Systems Using the GLUE Methodology." *Journal of Hydrology* 249 (1-4): 11–29. `https://doi.org/10.1016/S0022-1694(01)00421-8`.

Brady, Rebecca M., and Jphn S. Butler. 2021. "The Circle of Life: The Mathematics of Predator-Prey Relationships." *Frontiers for Young Minds* 9 (651131). `https://kids.frontiersin.org/articles/10.3389/frym.2021. 651131`.

Brown, Christopher. 2018. *Formula.tools: Programmatic Utilities for Manipulating Formulas, Expressions, Calls, Assignments and Other r Objects*. `https://github.com/decisionpatterns/formula.tools`.

Burnham, Kenneth P., and David R. Anderson. 2002. *Model Selection and Multimodel Inference*. New York, NY: Springer New York.

Carroll, Cameron Jewett. 2013. "Modeling Winter Severity and Harvest of Moose: Impacts of Nutrition and Predation." PhD thesis, Fairbanks, Alaska: University of Alaska Fairbanks.

Devore, Jay L., Kenneth N. Berk, and Matthew A. Carlton. 2021. *Modern Mathematical Statistics with Applications*. 3rd ed. Springer Texts in Statistics. Cham, Switzerland: Springer International Publishing. `https: //doi.org/10.1007/978-3-030-55156-8`.

Finnish Meteorological Institute. 2021. "Weather in Kuopio." https://en.ilmatieteenlaitos.fi/weather/kuopio.

Frank, Jacob W. 2021. "Snowshoe Hare." `https://commons.wikimedia.org/wi ki/File:Snowshoe/_Hare/_(6187109754).jpg`.

Gardiner, C W. 2004. *Handbook of Stochastic Methods for Physics, Chemistry, and the Natural Sciences.* 3rd ed. Berlin, Germany: Springer.

Gause, G. F. 1932. "Experimental Studies on the Struggle for Existence: I. Mixed Population of Two Species of Yeast." *Journal of Experimental Biology* 9 (4): 389–402.

Goulet, Vincent, Christophe Dutang, Martin Maechler, David Firth, Marina Shapira, and Michael Stadelmann. 2021. *Expm: Matrix Exponential, Log, Etc.* http://R-Forge.R-project.org/projects/expm/.

Keener, James P. 2021. *Biology in Time and Space: A Partial Differential Equation Modeling Approach.* Providence, Rhode Island: American Mathematical Society.

Keener, James, and James Sneyd, eds. 2009. *Mathematical Physiology.* New York, NY: Springer New York. https://doi.org/10.1007/978-0-387-75847-3.

Kermack, William Ogilvy, Anderson Gray McKendrick, and Gilbert Thomas Walker. 1927. "A Contribution to the Mathematical Theory of Epidemics." *Proceedings of the Royal Society of London. Series A, Containing Papers of a Mathematical and Physical Character* 115 (772): 700–721. https://doi.org/10.1098/rspa.1927.0118.

Kilby, Eric. 2012. "Canada Lynx." https://commons.wikimedia.org/wiki/File:Canada/_Lynx/_(8154273321).jpg.

King, Aaron A., and William M. Schaffer. 2001. "The Geometry of a Population Cycle: A Mechanistic Model of Snowshoe Hare Demography." *Ecology* 82 (3): 814–30. https://doi.org/10.2307/2680200.

Kuznetsov, Yuri. 2004. *Elements of Applied Bifurcation Theory.* 3rd ed. Springer.

Logan, J. David, and William Wolesensky. 2009. *Mathematical Methods in Biology.* 1st ed. Hoboken, N.J: Wiley.

Lotka, Alfred J. 1920. "Analytical Note on Certain Rhythmic Relations in Organic Systems." *Proceedings of the National Academy of Science* 6 (7): 410–15.

———. 1926. "Elements of Physical Biology." *Science Progress in the Twentieth Century (1919-1933)* 21 (82): 341–43.

MacLulich, D. A. 1937. *Fluctuations in the Numbers of the Varying Hare (Lepus Americanus).* Toronto, Canada: University of Toronto Press. https://doi.org/10.3138/9781487583064.

Mahaffy, Joseph. 2010. "Lotka-Volterra Models." https://jmahaffy.sdsu.edu/courses/f09/math636/lectures/lotka/qualde2.html.

Matthes, Jackie. 2021. "Bisc204_BioModeling." https://github.com/jhmatth
es/BISC204/_BioModeling.

Munz, Philip, Ioan Hudea, Joe Imad, and Robert J. Smith? 2009. "When Zom-
bies Attack!: Mathematical Modeling of an Outbreak of Zombie Infection."
In *Infectious Disease Modelling Research Progress*, 133–56. Happuage, New
York: Nova Science Publishers.

OpenStax, C. N. X. 2016. https://commons.wikimedia.org/wiki/File:
Figure/_45/_06/_01.jpg.

Pastor, John. 2008. *Mathematical Ecology of Populations and Ecosystems*. West
Sussex, United Kingdom: Wiley-Blackwell.

Perko, Lawrence. 2001. *Differential Equations and Dynamical Systems*. 3rd ed.
New York, New York: Springer.

R Core Team. 2021. *R: A Language and Environment for Statistical Computing*.
Vienna, Austria: R Foundation for Statistical Computing. https://www.R-
project.org/.

Richey, Matthew. 2010. "The Evolution of Markov Chain Monte Carlo Methods."
The American Mathematical Monthly 117 (5): 383–413.

RStudio Team. 2020. *RStudio: Integrated Development Environment for R*.
Manual. Boston, MA: RStudio, PBC. http://www.rstudio.com/.

Schloerke, Barret, Di Cook, Joseph Larmarange, Francois Briatte, Moritz
Marbach, Edwin Thoen, Amos Elberg, and Jason Crowley. 2021. *GGally:
Extension to Ggplot2*. https://CRAN.R-project.org/package=GGally.

Scholz, Gudrun, and Fritz Scholz. 2014. "First-Order Differential Equations
in Chemistry." *ChemTexts* 1 (1): 1. https://doi.org/10.1007/s40828-014-
0001-x.

Schwartz, G. 1978. "Estimating the Dimensions of a Model." *Annals of Statistics*
6 (2): 461–64.

Sethi, Suresh P. 1983. "Deterministic and Stochastic Optimization of a Dynamic
Advertising Model." *Optimal Control Applications and Methods* 4 (2): 179–
84. https://doi.org/10.1002/oca.1660010207.

Sinay, Laura, and Leon Sinay. 2006. "A Simple Mathematical Model for the
Effects of the Growth of Tourism on Environment." In *New Perspectives
and Values in World Tourism and Tourism Management in the Future,
20-26 November, 2006*. Alanya, Turkey.

Stenseth, Nils Chr., Wilhelm Falck, Ottar N. Bjørnstad, and Charles J. Krebs.
1997. "Population Regulation in Snowshoe Hare and Canadian Lynx:
Asymmetric Food Web Configurations Between Hare and Lynx." *Pro-*

ceedings of the National Academy of Sciences* 94 (10): 5147–52. `https://doi.org/10.1073/pnas.94.10.5147`.

Strogatz, Steven H. 2015. *Nonlinear Dynamics and Chaos: With Applications to Physics, Biology, Chemistry, and Engineering.* 2nd ed. Boulder, CO: CRC Press.

Thornley, John H. M., and Ian R. Johnson. 1990. *Plant and Crop Modelling: A Mathematical Approach to Plant and Crop Physiology.* Caldwell, New Jersey: Blackburn Press.

Volterra, Vito. 1926. "Fluctuations in the Abundance of a Species Considered Mathematically." *Nature* 118 (2972): 558–60. `https://doi.org/10.1038/118558a0`.

Wickham, Hadley, Mara Averick, Jennifer Bryan, Winston Chang, Lucy D'Agostino McGowan, Romain François, Garrett Grolemund, et al. 2019. "Welcome to the tidyverse." *Journal of Open Source Software* 4 (43): 1686. `https://doi.org/10.21105/joss.01686`.

Wickham, Hadley, and Garrett Grolemund. 2017. *R for Data Science: Import, Tidy, Transform, Visualize, and Model Data.* 1st edition. Sebastopol, CA: O'Reilly Media.

Wilson, Greg, D. A. Aruliah, C. Titus Brown, Neil P. Chue Hong, Matt Davis, Richard T. Guy, Steven H. D. Haddock, et al. 2014. "Best Practices for Scientific Computing." *PLoS Biol* 12 (1): e1001745. `https://doi.org/10.1371/journal.pbio.1001745`.

Zobitz, John. 2022. *Demodelr: Simulating Differential Equations with Data.*

Index